Making Sense of Heritability

In this book, Neven Sesardic defends the view that it is both possible and useful to measure the separate contributions of heredity and environment to the explanation of human psychological differences. He critically examines the view – very widely accepted by scientists, social scientists and philosophers of science – that heritability estimates have no causal implications and are devoid of any interest. In a series of clearly written chapters he introduces the reader to the problems and subjects the arguments to close philosophical scrutiny. His conclusion is that anti-heritability arguments are based on conceptual confusions and misunderstandings of behavior genetics. His book is a fresh, original, and compelling intervention in a very contentious debate.

NEVEN SESARDIC is Associate Professor in the Department of Philosophy at Lingnan University, Hong Kong. His areas of specialization are philosophy of biology, philosophy of science, and philosophy of mind.

CAMBRIDGE STUDIES IN PHILOSOPHY AND BIOLOGY

General Editor
Michael Ruse   *Florida State University*

Advisory Board
Michael Donoghue   *Yale University*
Jean Gayon   *University of Paris*
Jonathan Hodge   *University of Leeds*
Jane Maienschein   *Arizona State University*
Jesús Mosterín   *Instituto de Filosofía (Spanish Research Council)*
Elliott Sober   *University of Wisconsin*

Alfred I. Tauber   *The Immune Self: Theory or Metaphor?*
Elliott Sober   *From a Biological Point of View*
Robert Brandon   *Concepts and Methods in Evolutionary Biology*
Peter Godfrey-Smith   *Complexity and the Function of Mind in Nature*
William A. Rottschaefer   *The Biology and Psychology of Moral Agency*
Sahotra Sarkar   *Genetics and Reductionism*
Jean Gayon   *Darwinism's Struggle for Survival*
Jane Maienschein and Michael Ruse (eds.)   *Biology
and the Foundation of Ethics*
Jack Wilson   *Biological Individuality*
Richard Creath and Jane Maienschein (eds.)   *Biology
and Epistemology*
Alexander Rosenberg   *Darwinism in Philosophy, Social Science,
and Policy*
Peter Beurton, Raphael Falk, and Hans-Jörg Rheinberger (eds.)
The Concept of the Gene in Development and Evolution
David Hull   *Science and Selection*
James G. Lennox   *Aristotle's Philosophy of Biology*
Marc Ereshefsky   *The Poverty of the Linnaean Hierarchy*
Kim Sterelny   *The Evolution of Agency and Other Essays*
William S. Cooper   *The Evolution of Reason*
Peter McLaughlin   *What Functions Explain*

Making Sense of Heritability

NEVEN SESARDIC

Lingnan University, Hong Kong

CAMBRIDGE UNIVERSITY PRESS
Cambridge, New York, Melbourne, Madrid, Cape Town, Singapore,
São Paulo, Delhi, Dubai, Tokyo, Mexico City

Cambridge University Press
The Edinburgh Building, Cambridge CB2 8RU, UK

Published in the United States of America by Cambridge University Press, New York

www.cambridge.org
Information on this title: www.cambridge.org/9780521173339

First published 2005

A catalogue record for this publication is available from the British Library

Library of Congress Cataloguing in Publication Data
Sesardic, Neven, 1949–
Making sense of heritability
Neven Sesardic.
  p. cm. – (Cambridge studies in philosophy and biology)
Includes bibliographical references and index.
ISBN 0-521-82818-X
1. Nature and nurture. 2. Behavior genetics. 3. Genetic psychology.
4. Environmental psychology. I. Title. II. Series.
BF341.S47   2005
155.7–dc22   2005002871

ISBN 978-0-521-82818-5 Hardback
ISBN 978-0-521-17333-9 Paperback

The denial of genetically based psychological differences is the kind of sophisticated error normally accessible only to persons having Ph.D. degrees.
David Lykken

Contents

ix

Figures

Acknowledgments

While working on this book, I presented drafts of some chapters on different occasions and received useful feedback. I gave talks at Lingnan University, Konrad Lorenz Institute, the University of Rijeka, the University of Braunschweig, the University of Pittsburgh, Duke University, the University of Hannover, the University of Munich, the meeting of the British Society for Philosophy of Science in Glasgow (July 2002), the eighteenth biennial meeting of the Philosophy of Science Association in Milwaukee (November 2002), and the annual conference of the Australasian Association for Philosophy in Adelaide (July 2003). I also presented some of the material in Alex Rosenberg's graduate seminar in philosophy of biology at Duke University in April 2004, which was a valuable opportunity to test my ideas and have very stimulating discussions.

I am very grateful to my colleagues at Lingnan University for approving my request for a free semester at the crucial, final stage of my writing. My research assistant Mr. Chin Po San helped a lot with literature searches and with preparing the bibliography. The work on this book was supported by a grant from the Research Grants Council of the Hong Kong Special Administration Region, China (Project No. LU3005/01H), by an Alexander von Humboldt Foundation research grant, and by two Lingnan research grants (DA03A5 and DR04B6). I am grateful to the University of Chicago Press for permission to use the material from my two papers published in *Philosophy of Science* (© 2000 and 2003 by the Philosophy of Science Association).

I would like to thank Michael Ruse, the editor of the series Cambridge Studies in Philosophy and Biology, for his support for the project. I benefited a lot from comments on earlier drafts that I received from the following colleagues: Thomas J. Bouchard, Arthur Jensen, Noretta Koertge, Stefan Linquist, Paisley Livingston, John Loehlin, David Lubinski, Samir Okasha, Katie Pleasance, Robert Plomin, Nils Roll-Hansen, and Terry Sullivan. It should not be assumed that any of these people agree with the basic claims defended in this book.

Introduction

What is *heritability*? The heritability of a trait in a given population tells us what proportion of differences in that trait is due to genetic differences. It provides an answer to the main question in the nature–nurture controversy.

Heritability can be calculated both for physical and psychological characteristics but it generates controversy mainly in the context of human behavioral traits like intelligence, personality differences, criminality, etc. For this reason I will focus here only on discussions of heritability in psychology (i.e., human behavior genetics). Criticisms of heritability claims in this area are frequently based on a curious mixture of methodological objections and warnings about political motives and/or implications of this research. A systematic study that would try to disentangle all the argumentative threads that are often run together in this contentious debate was long overdue. So this book was simply waiting to be written.

I was introduced to the nature–nurture debate by reading Ned Block and Gerald Dworkin's well-known and widely cited anthology about the IQ controversy (Block & Dworkin 1976a). This collection of articles has long been the main source of information about the heredity–environment problem for a great number of scientists, philosophers, and other academics. It is not an exaggeration to say that the book has been the major influence on thinking about this question for many years. Like most readers, I also left the book with a feeling that hereditarianism (the view that IQ differences among individuals or groups are in substantial part due to genetic differences) is facing insuperable objections that strike at its very core.

There was something very satisfying, especially to philosophers, about the way hereditarianism was criticized there. A strong emphasis was on conceptual and methodological difficulties, and the central arguments against hereditarianism appeared to have full destructive force independently of empirical data, which are, as we know, both difficult to evaluate and inherently unpredictable.

So this looked like a philosopher's dream come true: a scientific issue with potentially dangerous political implications was defused not through an arduous exploration of the messy empirical material but by using a distinctly philosophical method of conceptual analysis and methodological criticism. It was especially gratifying that the undermined position was often associated with politically unacceptable views like racism, toleration of social injustice, etc. Besides, the defeat of that doctrine had a certain air of finality. It seemed to be the result of very general, *a priori* considerations, which, if correct, could not be reversed by "unpleasant" discoveries in the future.

But very soon I started having second thoughts about Block and Dworkin's collection. The reasons are worth explaining in some detail I think, because the book is still having a considerable impact, especially on discussions in philosophy of science.

First, some of the arguments against hereditarianism presented there were just *too* successful. The refutations looked so utterly simple, elegant, and conclusive that it made me wonder whether competent scientists could have really defended a position that was so manifestly indefensible. Something was very odd about the whole situation.

Second, the selection of texts in the anthology was extremely biased. Throughout the book the view that IQ is important and heritable was exposed to an unrelenting barrage of attacks, but in its more than 500 pages there was not a single text, however short, that would help the reader understand the basic rationale for that view, before its demolition began. The book started "in the attack mode," with a series of articles from the 1920s, in which a famous journalist with no obvious expertise in psychology criticized IQ tests and the idea that IQ differences are heritable. Here is a brief excerpt that gives the flavor of Walter Lippmann's polemical style:

> If, for example, the impression takes root that these tests really measure intelligence, that they constitute a sort of last judgment on the child's capacity, that they reveal "scientifically" his predestined ability, then it would be a thousand times better if all the intelligence testers and all their questionnaires were sunk without warning in the Sargasso Sea. (Block & Dworkin 1976a: 19)

2

In the anthology, Lippmann's combative articles were not preceded by any context-setting piece in which at least the bare essentials of psychometric theory would be explained in a neutral or sympathetic way. Similarly, two sections devoted to Arthur Jensen's and Richard Herrnstein's views on IQ and its heritability again started in a paradoxical way, with Lewontin's and Chomsky's attacks, without the reader being given any opportunity to learn beforehand about Jensen's and Herrnstein's theories from an impartial source, let alone from their own writings. The only firmly hereditarian texts included in the book were two brief responses to criticisms, but they were also immediately neutralized by further rejoinders.

Although this way of introducing the debate (presenting objections to a theory, without first presenting the theory itself) is so openly prejudiced and illogical, no one ever drew attention to it nor suggested that, because of the obvious bias, the anthology should be supplemented with an "antidote" reading list. Oddly enough, Peter Medawar (1977a) even praised the editors for their judicious selection of texts. Philip Kitcher similarly said that he "learned much" from the book, but he also apparently didn't notice that he could have learned from it only about one side of the debate. Worse still, he mistakenly claimed (Kitcher 2001a: 209) that Jensen's controversial article (Jensen 1969b), the main target of criticisms, was reprinted in Block and Dworkin's anthology. Kitcher's mistake is somewhat surprising because Jensen's book-size article, if included, would have completely dominated the anthology: it would have taken up nearly one-third of it.

Block and Dworkin themselves gave a very veiled hint of their own bias in the extremely short introduction. They wrote: "The title of the reader [*Critical Readings*] is accurate. We do not attempt to present 'all sides of the issue.' We have brought together the best of the critical literature" (Block & Dworkin 1976a: xi). Notice the scare quotes, which seem to suggest that the goal of presenting all sides of the issue in an anthology is somehow problematic or not really desirable.

Furthermore, the word "critical" is ambiguous. First, "critical" can mean "critical of one specific point of view." In that sense, choosing "the best of the critical literature" would not be a good editorial decision because it simply means "a very biased presentation of the literature." Second, "critical" can mean, according to *Webster*, "characterized by careful evaluation and judgment (example: 'a critical reading')," in which case it is indeed highly commendable for an anthology. But clearly in that sense, their selection did *not* contain "the best of the critical literature," because one-half of the best literature was virtually not represented.

Block and Dworkin traded on this ambiguity. They encouraged the reader to understand "critical" in the second sense by giving the title "Critical Readings," which even the dictionary links to the second sense, and by failing to reveal in the introduction the all-important fact about their book that supports the first-sense reading, namely, that the literature they selected was highly skewed *against* one position in the debate. Apart from Jensen's and Herrnstein's short texts, each of which was placed between the hammer and anvil of two rebuttals, practically all other texts were critical of psychometrics and/or hereditarianism, in one way or another.

But the first sense of "critical" had a useful function too. It gave Block and Dworkin a convenient fallback position in case they were criticized for not declaring their undeniable editorial bias. No other word could have served the purpose so well.

Despite its glaring tendentiousness, the book has been taken by many people as the *only* source of information about the controversy. Moreover, it is still being widely quoted and used as a recommended reading, with no warning about its bias or the need to balance this one-sided intellectual diet with texts in which hereditarianism is defended, or at least treated as a serious theoretical option. Although other, similarly biased books (Gould 1981; Rose et al. 1984) are also often uncritically advertised as good introductions to the nature–nurture debate, Block and Dworkin's anthology played a special role in the emergence of the anti-hereditarian consensus in mainstream philosophy of science.[1] It contained two articles written by Richard Lewontin, which quickly won almost universal acceptance and which are still regarded as disclosing the allegedly fatal methodological shortcomings of hereditarian thinking. When it comes to discussions about measuring the strength of genetic influences on psychological differences (whether among individuals or among groups), the importance of heritability is immediately dismissed just by quoting Lewontin's arguments, which are typically treated as the last word on the topic. (For references, see particularly chapters 2 and 4.) As Stephen Downes says in his article on heritability for the *Stanford Encyclopedia of Philosophy*, "the current consensus among philosophers of biology is that heritability analyses are misleading about the genetic causes of human traits . . . The consensus

---

[1] Paul Meehl said in 1970 that at that time most philosophers received "the environmentalist brainwashing" from their undergraduate social science classes (Meehl 1970: 393–394). The situation changed in the mid-1970s: philosophers started to be exposed to the anti-hereditarian indoctrination in their own departments.

among philosophers of biology is that broad heritability measures are uninformative" (Downes 2004: 7, 13).

To the best of my knowledge, there have been only three attempts to crack the philosophical consensus, all of them unsuccessful. First, even before the anti-heritability view hardened into a dogma, Peter Urbach published an article (Urbach 1974) in which he argued, using the Lakatosian approach to philosophy of science, that hereditarianism is a more promising research program than environmentalism. Only two brief responses appeared immediately after publication, but Urbach's contribution made no long-term impact. Second, Michael Levin raised objections to the "received view" on heritability in his thought-provoking book about racial differences (1997: 91–102; see also Levin 1994 and Hocutt & Levin 1999). However, his work was largely ignored in the philosophical literature, at least in part, I suspect, because Levin also defended heretical political conclusions about race in such an open and blunt manner that his views were simply regarded as too shocking to deserve comment. Last (and least), I myself published three papers (1993; 2000; 2003) in the attempt to soften the anti-hereditarian consensus, but without much success. I try again here.

The peculiarity of the philosophical consensus is that it has been preserved for thirty years, with very little interest among philosophers of biology in taking a closer look into the propulsive field of behavior genetics and seeing how heritability analyses look from the perspective of real empirical research. In a way, this is understandable. For if a given research orientation is methodologically doomed, why should one waste time on examining an enterprise that one "knows" must inevitably end in failure? Those who remain under the spell of Lewontin's dismissal of hereditarianism can explain a persistent interest in heritability in no other way than as "an unfortunate tic from which [behavior geneticists] cannot free themselves" (Kitcher 2001b: 413). Indeed, if you go with the consensus and strongly believe that hereditarianism is intellectually bankrupt and demonstrably so, but nevertheless see a lot of people still expressing enthusiasm for that very view, you will soon despair and start asking yourself: "Hey, how come this totally discredited position is having a revival? How many times do we have to kill this already dead beast?" You will have a feeling that you are fighting a "vampire" (Griffiths 2002a) that haunts people even after being killed (refuted), that you are confronting "the ghost of dichotomous views of development" (Gray 1992: 172), that the issue is "periodically disinterred from its well-deserved grave" (Johnston 1987: 150), that you are "battling the undead" (Kitcher 2001b), that "the corpse will not stay dead" (Paul 1998: 83), that the dead will not "remain in their coffins" (Kitcher 1984: 9), and you will look for the final "stake-in-the-heart" move (Oyama 2000c: 31; Oyama et al. 2001: 1) that will finish off this theoretical monster.[2]

Griffiths, Kitcher, and Oyama disagree about many things, as do others who belong to the consensus. The consensus is not monolithic. Yet virtually all of them agree that Lewontin's canonical arguments conclusively demonstrated methodological limitations that severely reduce the theoretical importance of any heritability estimates obtained with non-experimental methods (which is the only available approach in the context of human psychology).

The alternative that these authors never seriously tried is to question the consensus and open their minds to the possibility that research in behavior genetics (including the measurement of heritability) actually makes more sense than they think.

Curiously, those on the other side of the debate did not show much interest, either, in meeting the opponents on their ground. Behavior geneticists typically respond to very general methodological objections in an ad hoc manner, often in the course of reporting about their own empirical research, and rarely address these global criticisms head-on or in a systematic way. Although they are very responsive to critical comments that can help improve the design of their studies or suggest new ways of hypothesis testing, there is a distinct impatience with those highly conceptual arguments that produced the philosophical consensus and that have no immediate relevance for daily scientific practice. Thomas Bouchard explains well why many people think that this type of argument does not deserve serious consideration:

> We need not dwell on these arguments for long. If they were at all persuasive there would be little, if any, need to attack the evidence underlying the hereditarian viewpoint. It would fall of its own weight. The massive, and vituperous, attacks on hereditarian findings clearly signal how seriously the environmental program is challenged by this evidence. (Bouchard 1987: 58)

This attitude explains why in the literature about heritability, even after all these years, there is still no systematic study in which methodological arguments against hereditarianism would be given full attention and examined in critical spirit. This lacuna harms the hereditarian case. I think

---

[2] By the way, the use of the metaphor of beating a dead horse that is not really dead to make a point about the nature–nurture debate is usually attributed to Susan Oyama, but it actually originates from Donald Hebb (1980: 70).

that Bouchard is seriously mistaken when he says that there is "no need to dwell on these arguments for long," and that it is enough just to focus on collecting empirical evidence and hope that it will eventually sway public opinion in favor of hereditarianism. The problem with his view is that the persuasiveness of empirical evidence for a given hypothesis crucially depends on the perceived methodological soundness of that hypothesis. If someone believes that hypothesis H has inherent and ineradicable methodological shortcomings, then further accumulation of empirical evidence will not make that person change his mind about H. If H is thought to have a "construction defect," piling up new evidence will look like pouring water into a bucket that has a hole in it.

Of course, Bouchard would be justified in not worrying too much about these global methodological criticisms if the only people who made a fuss over them were philosophers of science. Even with this unfriendly stance becoming a consensus in philosophy of science, scientists might still remain unimpressed because many of them would probably be sympathetic to James Watson's claim: "I do not like to suffer at all from what I call the German disease, an interest in philosophy" (Watson 1986: 19).

But the situation is worse than that. Methodological doubts about the usefulness of partitioning causal contributions of heredity and environment have already spread outside of philosophy to many areas of social science. Most importantly, however, the same views that are part of the anti-hereditarian consensus in philosophy of science were presented in a watered-down form to the general audience, and they carried a lot of conviction there. Popularizers of science bombarded the public with all sorts of arguments that all pointed in the same direction, such as that the attempts to measure heritability are unscientific, naive, crude, obviously wrong, etc. To support these criticisms they made different points on different occasions: that genetic and environmental influences are inseparable in development; or that the impact of genotype cannot be measured because it depends on the environment; or that genetic causes often produce their effects through environmental pathways, and that classifying them as genetic is therefore bound to create confusion; or that twin studies, an important source of heritability estimates, are based on an uncritically accepted assumption that is probably false; or that the heritability of a trait says nothing about its malleability, and is hence uninteresting; and so on.

The reader is forced into submission by the sheer abundance and diversity of objections. Even if he finds some of these objections unpersuasive, there will be others to which he doesn't have a ready response, and so he will in the end join the camp of heritability skeptics. There is nothing wrong in principle with this kind of approach, where a given view is disputed by pulling together very different strands of argument against it. For it may well be that although each of these arguments has some problems, they nevertheless collectively build a formidable case against the criticized position. But then again, it may be that the situation is different and that, taken individually, these arguments are so weak that even when they are harnessed together, they still do not amount to something very impressive. Simply, we cannot know what the situation is before we undertake a careful and detailed examination.

An additional reason why Bouchard is wrong that there is not much need to discuss purely methodological criticisms is that, unless they are addressed and opposed, these criticisms tend to mix with political considerations to form a deadly anti-hereditarian combination that is extremely difficult to dislodge by mere argument. Let's face it, many people *wish* that hereditarianism not be true because the hypothesis of inherited psychological differences doesn't sit well with their deeply ingrained beliefs about human equality and "perfectibility of man." In particular, the idea that groups with physically recognizable characteristics differ in the average genetic potential for intellectual achievement offends political sensibilities so much that there is a positive desire to find reasons against it.[3] In fact, both hereditarians and environmentalists recognize that there is widespread and deep-seated repugnance toward the heritability of psychological traits. This is one of the very few things about which even people as politically opposed as Richard Herrnstein and Bill Clinton can agree:

> Biological determinism runs "against the American grain." (Herrnstein 1973: 5)

> I disagree with the proposition that there are inherent, racially based differences in the capacity of the American people to reach their full potential; I just don't agree with that. It goes against our entire history and our whole tradition. (Clinton's comments on *The Bell Curve*, at the White House press conference, September 21, 1994)

Therefore, in the inhospitable situation where hereditarianism is resisted because it is perceived (rightly or wrongly) to threaten cherished

---

[3] I concede that some people's wishful thinking may go in the opposite direction, in the sense that they would revel in the discovery of such genetic differences. Yet it is undeniable that the first tendency is immensely more prevalent among scientists, academics, journalists, public intellectuals, politicians, and other opinion makers.

political values, and where even highly dubious methodological arguments are welcome if they are the best way to protect these values, hereditarians are well advised to put these arguments under the magnifying glass, dissect them, and highlight problematic inferences, questionable assumptions, or outright fallacies.

In fact, this is even more of a task for philosophers of science, who are after all specially trained to be attentive precisely to methodological and conceptual aspects of scientific controversies. But for some reason, in this debate philosophers have displayed a surprising lack of intellectual curiosity and analytical acuity. They hastily accepted some general anti-hereditarian arguments that possessed only superficial plausibility. Soon these arguments, without being exposed to adequate critical scrutiny, rigidified into a philosophical consensus. The paradigm was established and ruled for decades, not because of its theoretical advantages but because its problematic sides went unnoticed. Easily anticipated objections were not considered at all, obvious alternatives were not explored, and gross misinterpretations created the illusion of an easy victory. To make things worse, and quite unusually for this field otherwise known for its high intellectual standards, in this small segment of philosophy of science even prominent scholars are often poorly informed about basic scientific facts in the very domain of their philosophical explorations.

This book is an attempt to dispel a number of obstinate misconceptions and pseudo-arguments about heritability that have dominated the intellectual scene for too long and that have diverted scientific and philosophical attention from the really interesting issues. Is such a predominantly negative goal worth the effort? I think it is. For if John Locke could have thought in his time that "it is ambition enough to be employed as an under-labourer in clearing the ground a little, and removing some of the rubbish that lies in the way to knowledge" (Locke 1959: 14), wouldn't it be arrogant on my part to declare that this kind of task is too small for me or not rewarding enough?

At this point I am afraid I may lose some of my scientific readers. Remembering Steven Weinberg's statement that the insights of philosophers have occasionally benefited scientists, "but generally in a negative fashion – by protecting them from the preconceptions of other philosophers" (Weinberg 1993: 107), they might conclude that it is best just to avoid reading any philosophy (including this book), and that in this way they will neither contract preconceptions nor need protection from them. But the problem is that the preconceptions discussed here do not originate from a philosophical armchair. Scientists should be aware that to a great extent these preconceptions come from some of their own. Philosophers of science uncritically accepted these seductive but ultimately fallacious arguments from scientists, repackaged them a little, and then fed them back to the scientific community, which often took them very seriously. Bad science was mistaken for good philosophy.

Here is a brief overview of the content of the book. In chapter 1, I consider some criticisms of the nature–nurture dichotomy and some elementary confusions about heritability. Chapter 2 addresses the complaint that heritability estimates are devoid of causal implications because of statistical interaction between genotypes and environments. (I discussed this topic in my first published paper on heritability [Sesardic 1993], but the argument presented here is completely reworked, much expanded, and, I hope, more convincing.) The topic of chapter 3 is the question whether the possibility of genetic effects being mediated by environmental influences shows, as many scientists and philosophers believe, that heritability is not a very helpful causal or explanatory notion. In chapter 4, the focus is on between-group heritability and its relation to within-group heritability. The political temperature rises in this context because the between-group heritability obviously becomes an issue with respect to the racial gap in IQ. In chapter 5, I discuss whether a trait's being heritable tells us anything about the modifiability (or malleability) of that trait. Finally, chapter 6 contains my thoughts about connections between science and politics, and about possible political implications that would follow from the high heritability of some psychological traits.

1

The nature–nurture debate: a premature burial?

> The nature–nurture problem is nevertheless far from meaningless. Asking right questions is, in science, often a large step toward obtaining the right answer.
>
> Theodosius Dobzhansky

Heritability is basically a measure of the strength of genetic influence on phenotypic differences.[1] The emphasis is on the word "differences." It was Francis Galton who initiated a systematic study of human variation in the nineteenth century. He chided statisticians of his time for being only interested in the mean values and never in differences:

> It is difficult to understand why statisticians commonly limit their inquiries to Averages and do not revel in more comprehensive views. Their souls seem as dull to the charm of variety as that of the native of one of our flat English counties, whose retrospect of Switzerland was that, if its mountains could be thrown into its lakes, two nuisances would be got rid of at once. (Galton 1889: 62)

Heritability is usually defined as the proportion of phenotypic variation that is due to genetic differences. So giving a particular value to a heritability estimate of a given trait is solving the nature–nurture equation in that specific context. There is no *general* answer to the nature–nurture question. As J. B. S. Haldane said: "The important point is to realize that the question of the relative importance of nature and nurture has no

---

[1] "Phenotype" refers to all observable characteristics of an organism, as contrasted with its genetic constitution, or its "genotype."

11

general answer, but that it has a very large number of particular answers" (Haldane 1938: 34). Dobzhansky concurs: "There is not one nature–nurture problem but many" (Dobzhansky 1956: 21).

Before going into a more detailed discussion of heritability, it is interesting to note that there is a widespread tendency to see the whole nature–nurture question as the paradigm of extremely crude and unsophisticated thinking. In contrast to Galton, who believed that the phrase "nature and nurture" is "a convenient jingle of words" (Galton 1874: 12) that delineates a fascinating scientific problem, it has been often suggested that quantifying the contributions of these two factors is meaningless, and moreover manifestly so. The standard argument for dismissing the controversy is that *both* nature and nurture are necessary for the life of any organism, and that for this reason the idea of measuring the relative importance of the two factors must be the result of serious confusion. In defending this diagnosis, Friedrich Hayek invokes the authority of the biologist Gavin de Beer:

> On the whole we must probably conclude, as does Sir Gavin de Beer . . . that the old controversy between "Nature" and "Nurture" ought to be allowed to die because "it is necessary to regard both nature and nurture as cooperating, without our being able to say in any one case exactly how much has been contributed by either." (Hayek 1978: 294)

Patrick Bateson's judgment is even harsher. After briefly reviewing the development of the nature–nurture discussion, he says:

> Any scientific investigation of the origins of human behavioral differences eventually arrives at a conclusion that most non-scientists would probably have reached after only a few seconds' thought. Genes and environment both matter. The more subtle question about how much each of them matters defies an easy answer. There is no simple formula to solve this conundrum, and the problem needs to be tackled differently. (Bateson & Martin 2000: 61)

It is hard to imagine a more damning criticism of a scientific research area than the blunt claim that most non-scientists would probably have reached the right conclusion about the matter in question "after only a few seconds' thought." In fact, it is safe to argue, contra Bateson, that since it would be astonishing if scientists did not recognize such a trivial answer in their own domain of research, it must be that the real issues lie elsewhere, and that the debate simply could not have been *so patently* meaningless. Also, the fact that there is indeed no "easy answer" in dividing the contributions of nature and nurture and that there is "no simple formula to solve this conundrum" does not immediately show that "the problem needs to be tackled differently." It may be that the question is neither so trivial as to be unworthy of investigation nor so intractable that a completely different approach is necessary.

Joseph LeDoux also treats the nature–nurture controversy as an easy target for criticism in a short, tangential remark:

> One of the most important contributions of modern neuroscience has been to show that the nature/nurture debate operates around a false dichotomy: the assumption that biology, on one hand, and lived experience, on the other, affect us in fundamentally different ways. Research has shown that not only do nature and nurture each contribute (in disputable proportions) to who we are, but also that they speak the same language. Both achieve their effects by altering the synaptic organization of the brain. (LeDoux 1998)

Two critical comments. First, surely the fact that both nature and nurture achieve their effects mostly by influencing the brain was widely known before modern neuroscience developed. Second, even though both kinds of causes do speak the same language ("brainese"), it may still be true that they tend to affect us in sufficiently different ways as to justify keeping them analytically apart in some contexts.

Therefore, let us not too hastily agree that the nature–nurture debate is a "pseudoquestion" (Lewontin 1976a: 181; Hirsch 1976: 171), "an illogical construct" (Daniels et al. 1997: 64), "dead" (Ridley 2003a: 280), "a naive question" (Mazur & Robertson 1972: 83), "an incorrectly phrased question" (Kagan & Snidman 2004: 29), "faulty" (Gottlieb 2001: 402), "a dead issue" (Anastasi 1958: 197), "an unnecessary detour" (Sternberg & Grigorenko 1999: 536), "an unnecessary debate" (Khoury & Thornburg 2001), "a graveyard of rotting doctrines" (Kitcher 1996: 250), "a false dichotomy" (Ehrlich 2000), "the false, dichotomous model" (Gould 1995), "dull query" (Jones 1994a: 226), "the foolish question" and "a fool's errand" (Meaney 2001: 50–51), "sterile" (Wahlsten 1990b: 150; Oyama 1992: 228), "counterproductive" (Keller 2001: 299), "deader than a doornail" (Schneider 2003: 137), "rubbish" (Jacquard 1985: 51), or that it should be "silently carried to its grave" (De Waal 1999: 99), etc. It is true that, interpreted in some ways, the issue is indeed "unworthy of further consideration" (Anastasi 1958: 197), but the sillier these versions are, the less likely it is that they really connect with the focal scientific debates of the past.

It is simply a historical distortion to present the heredity–environment controversy as having been resolved by the "insight" that *both* genes and environment matter for development. No one ever doubted this obvious and rather unilluminating truth. It is not surprising, therefore, that when criticizing the idea that nature and nurture are independent, separable sources of influence, each of which can act "in isolation from the other," Lerner (1986: 83–84) can list the names of those who disputed that claim (Anastasi, Lehrman, Hebb, Schneirla) but that he cannot cite a single source where such a patently absurd idea has been defended. Also, when Lewontin says that "it was supposed that the phenotype of an individual could be the result of *either* environment *or* genotype, whereas we understand the phenotype to be the result of *both*" (Lewontin 1976a: 181), no reference is given for the preposterous monocausal view, simply because no such references exist.

The suggestion that there are (or were) some people, usually called "genetic determinists," who believe that genes are "self-sufficient causes" of behavior (Keller 1994: 118; Buchanan et al. 2002: 23), that "genes alone can produce traits" (Moore 2001: 13), that "only genes matter" (Sterelny & Griffiths 1999: 59), that "we are completely determined by our genes" (Millstein 2002: 233), or that "the development of an organism is determined solely by genetic factors" (Baker 2004: 17) is ludicrous. Everybody has always known that, as David Lykken puts it, "without environmental input, your genome would have created nothing more than a damp spot on the carpet" (1995: 85). Therefore, if "genetic determinism" is defined as the doctrine that an organism's phenotype is determined by genotype alone,[2] it soon becomes obvious that "every first-year biology student knows [that it is false]" (Antony 2000: 17), and then surely the question must arise: why should such a silly belief be dignified with a name, or debated at all?

As there is literally no one who ever subscribed to genetic determinism,[3] the meaning of this doctrine obviously cannot be inferred from the texts of its (non-existent) proponents. But then the only way to reconstruct the thesis of genetic determinism is, oddly enough, to distill it from

---

[2] "[T]he phrase 'genetic determinism' would, strictly speaking, mean that every event has a genetic cause sufficient for that event's occurring" (Wachbroit 2001: 31).

[3] One of the rare philosophers who said the right thing about genetic determinism is Robert Richardson: "I know of no historical writer of merit or competent contemporary theorist that has embraced a genetic determinism according to which specific behaviors or behavioral types are expressed despite, or even independently of, environmental influences. The only question having any significance is one of the relative contributions of genetic and environmental factors" (Richardson 1980: 481).

the texts of its *opponents*: "in *denying* the thesis of genetic determinism various researchers and other writers make various sorts of claims, and *from these claims some ideas about the form of the thesis they are arguing against can be gleaned*" (Kaplan 2000: 10, italics supplied). A straw man in the making?

To make things worse, the new and much discussed Developmental Systems Theory (DST) is sometimes described as "rejecting the idea that genes *or* environments can create traits by themselves" (Moore 2001: 8), and this trivial claim is even occasionally represented as the "essence" of the new perspective (Moore 2001: 61). Similarly, it is claimed that the *main* conclusion of the developmentalist argument is that "both genes and environment make essential developmental contributions to all aspects of our behavior" (Johnston 2003), or that it is "at the heart of developmental analysis" that "genes in themselves cannot cause development" (Gottlieb 1992: 162), or that from the DST perspective, a gene (or an environmental influence) does not "singly" determine the phenotype (Moss 2003: 115). But if this is really its crucial message, then one might as well read the abbreviation DST as "Definitely Superfluous Theory."[4]

The following quotation from a very early overview of heritability research shows that reminding behavior geneticists that both genes and environments matter is like carrying coals to Newcastle:

> if there is no environment, no organism can develop to display any pheno-type whatsoever. Likewise, without a genetic constitution, there will be no organism. It is clear that a question asking if a trait is due to heredity or to environment is nonsense. Without both, there would be no trait at all. A meaningful answer, however, may be sought to a question concerning the relative contributions of genetic differences and environmental differences to the variability of a characteristic. (McClearn 1964: 167)

To go even further into the past, Francis Galton, whom many regard as "the founder of behavior genetics" (McGuffin 2000: 243), was clearly aware of that truism. He stated it just in passing, having regarded it already

---

[4] I don't want to say that the whole DST consists of trivial claims. But I do want to say, first, that the dismissal of heritability in DST is just an uncritical repetition of Lewontin's arguments, and, second, that sometimes it is really hard to see who is the addressee of their criticisms of hereditarianism. Example: "Nature has no existence prior to or separate from the concrete living organism in its concrete, often living, surroundings: no Platonic ideals here, no underlying reality more basic than the being itself, no instruction manuals or little engineers in the cell nucleus" (Oyama 2002: 164). Who is being attacked here? Who thinks that nature has existence prior to or separate from the concrete living organism?

at that time as too self-evident to require further elaboration: "*It is needless to insist* that neither [nature nor nurture] is self-sufficient" (Galton 1874: 12, italics added).

1.2 NO FAIR HEARING FOR FRANCIS GALTON

Speaking of Galton, "that versatile and somewhat eccentric man of genius" (Fisher 1959: 1), some criticisms of his work additionally illustrate the widespread tendency among contemporary scholars to misrepresent the hereditarian theorizing from its historical beginnings, and make it appear as overly crude, simplistic, or downright silly. For example, discussing Galton's famous attempt to determine the influence of heredity on achieving eminence, David Moore finds a fatal flaw: Galton "ignored what is now obvious to us," i.e., that the children of eminent people shared not just heredity with their parents but also favorable environments. According to Moore, this oversight shows that Galton "had a preexisting bias that sometimes prevented him from seeing the contributions that environmental factors can make to the appearance of our traits" (Moore 2001: 35–36).

In reality, Galton was too much of a statistician to be unaware of such an obvious need to consider an alternative explanation. At the beginning of *Hereditary Genius* he writes: "I must not compare the sons of eminent men with those of non-eminent, *because much of which I should ascribe to breed, others might ascribe to parental encouragement and example*" (Galton 1892: 37, italics supplied). It was precisely for this reason that he undertook to compare the proportion of eminent people among the biological children of eminent people with adopted sons of Popes and other dignitaries of the Roman Catholic Church, in the attempt to disentangle possibly confounding influences from heredity and environment. In his own words: "If social help is really of the highest importance, the nephews [adopted children] of the Popes will attain eminence as frequently, or nearly so, as the sons of other eminent men; otherwise, they will not" (Galton 1892: 42). And the data clearly indicated a significant difference between the two groups, leading Galton to conclude that the environmental explanation could not be the whole story. This does not prove that Galton was right, of course, but it does show that his argument was immensely more sophisticated than Moore pictured it.

Notice also how quickly Moore resorts to the attribution of "a preexisting bias." Apparently, the conviction that early hereditarianism must

16

have been appallingly prejudiced and unrefined is acquired through social osmosis, with the result that the view is dismissed on invented grounds and with no acquaintance with the relevant texts. (Although Moore discusses Galton's views extensively, there are no references to Galton's works in his text or bibliography.)

The ill will toward the father of eugenics – Peter Medawar referred to him as "wicked old Sir Francis Galton" (Medawar 1977b) – seems to be so strong that even his indisputable achievements tend to be recognized only grudgingly.[5] For example, Moore describes Galton not as a person who invented regression analysis, but as someone who "stumbled" onto it (2001: 37). The word "stumble" is singularly inappropriate here because in reality Galton not only introduced the concept of regression, but also slowly and single-handedly refined it over a period of twenty years in what a historian of statistics called "one of the grand triumphs of the history of science" (Stigler 1999: 177).

Galton suffered a similar fate at the hands of geneticist Steve Jones:

> Galton, in *Hereditary Genius*, went to great lengths to show that talent runs in families and was coded into our biology. Oddly enough, he never pointed out that more than half his "geniuses" turned up in families with no history of distinction at all. Such was the power of prejudice that he concentrated only on those who supported his hereditarian views. (Jones 1994a: 225)

Again, an alleged methodological gaffe is unearthed, with an imputation of prejudice promptly following. But Jones's methodological criticism is both factually incorrect and entirely irrelevant. First, Galton *did* point out that among poets, musicians, painters, divines, and scholars, more than half of the illustrious men came from families with no history of distinction (Galton 1892: 228, 239, 249, 274, 283, 300). Furthermore, summing up the information for all the professions that he considered, Galton concluded from his tabulated data that, cumulatively, "exactly one-half of the illustrious men have one or more eminent relations" (Galton 1892:

---

[5] In his diatribe against "the unnatural science" of IQ and its heritability, Medawar says that "God alone knows how [Lewis Terman] estimated Galton's IQ as 200" (1977b), implying that such an estimate could have been based only on some very weird and ridiculous speculation. But in fact Terman's method was very simple and straightforward (Terman 1917: 209). On the basis of Karl Pearson's biography of Galton, which gave ample evidence of Galton's extraordinary precocity, Terman concluded that between the age of three and eight years Galton's mental age (MA) was approximately double his actual age (CA). Now if in Galton's case $MA = 2 \times CA$, then in accordance with the then standard practice of calculating IQ, the equation "$IQ = (MA/CA) \times 100$" immediately gives the value of his IQ: $(2 \times CA)/CA \times 100 = 200$.

322), from which it deductively follows that one-half of the illustrious men did *not* have any eminent relations. Second, it is unclear why Jones thinks that the fact of more than half of geniuses turning up in families with no distinction would create a problem for Galton's hereditarianism. Since Galton's illustrious men were selected in such a way that their ratio to the general population was 1 to 4,000, the hereditarian explanation could easily accommodate the proportion of eminent people from no-distinction families significantly higher than 50 percent.[6]

It may seem surprising that Jones dismissed the views of the founder of his own laboratory (Galton Laboratory, University College London) in such a manner. But then again this should perhaps not be so surprising. One can hardly be expected to study seriously the work of a man whom one happens to call publicly "Victorian racist swine" – the way Jones referred to Galton in an interview (Grove 1991). Also, in Jones's book *Genetics for Beginners* (Jones & Van Loon 1993: 169), Galton is pictured in a Nazi uniform, with a swastika on his sleeve.

1.3 HERITABILITY 101

Although both genes and environmental influences are absolutely *necessary* for the development (and even just for the continuing existence) of organisms, this does not mean that the importance of these two factors must be the same. One measure of that difference is heritability, usually symbolized by "h^2."

Let me illustrate how the concept works in a very simple situation. Consider a population that consists of organisms with three different genotypes (G_1, G_2, and G_3) that are randomly distributed in three different environments (E_1, E_2, and E_3). Table 1.1 represents the phenotypic values in a given trait for all nine G–E combinations, and phenotypic means of particular genotypes (G-means) as well as phenotypic means of specific environments (E-means).

Even without any statistical analysis it is easy to see that in this population genetic effects on phenotype are considerably stronger than the environmental ones. Keeping the environment fixed, a change from G_1

---

[6] The same criticism is leveled against Galton by Joseph L. Graves, who in the end concludes that Galton was "an intellectual mediocrity, a sham, and a villain" (Graves 2001: 100). In a brief review of Graves's book, Loring Brace (professor of anthropology at the University of Michigan) praises the author for these very words and says that "his demonstration of the truth of this description is worth the price of the book" (Brace 2001).

Table 1.1

	E_1	E_2	E_3	G-means
G_1	1	2	3	2
G_2	4	5	6	5
G_3	7	8	9	8
E-means	4	5	6	Grand mean: 5

to G_2, or from G_2 to G_3 always results in the phenotypic increase of three points, whereas the corresponding transition from one environment to the other produces the phenotypic rise of only one point. The so-called "main effect" of genotype is visible in the last column, and the main effect of environment is seen in the last row. The fact that a genotypic change affects the phenotype more strongly than an environmental change is manifest in Figure 1.1.

An alternative way to describe the data in Table 1.1 is by using the concept of variance. The total phenotypic variance (V_P) in the population is calculated by first squaring every individual deviation from the overall phenotypic mean (5), then adding up all these values, and finally dividing the result by the number of individual cases (9). When we do this, we obtain the result: 6.66. Now this total phenotypic variance can be decomposed into two parts: the one due to genetic differences (genetic variance, or V_G) and the one caused by environmental differences (environmental variance, or V_E). To calculate V_G, we square the deviations of three genotypic means from the grand mean, add them up, and divide by 3: $(3^2 + 0^2 + 3^2)/3 = 6$. So, the genetic part of phenotypic variance (or

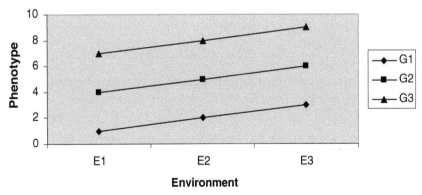

Figure 1.1 A simple illustration of heritability.

genetic variance, in short) is 6. *Mutatis mutandis*, we do the same thing for the environmental part, and we get the value for environmental variance (V_E): $(1^2 + 0^2 + 1^2)/3 = 0.66$.

Now we are ready to calculate the heritability, i.e., the proportion of genetic variance in the total phenotypic variation. We divide genetic variance V_G by total variance V_P: $6/6.66 = 0.9$. The heritability (h^2) is 90 percent. (Let me stress that this is an unrealistically high value of heritability for most real-life examples.) The contribution of environment to phenotypic variance (sometimes called environmentality, or e^2) is 0.1 ($0.66/6.66$), or 10 percent.

All this connects directly to what was discussed before:

> The partitioning of the variance into its components allows us to estimate the relative importance of the various determinants of the phenotype, in particular the role of heredity versus environment, or nature and nurture ...
> The relative importance of heredity in determining phenotypic values is called the *heritability* of the character. (Falconer 1989: 125–126)

It should be stressed again that my illustration with three genotypes and three environments is an unrealistic simplification, introduced just for expository convenience at this early stage.[7] The real-life examples get much more complicated, basically in two ways.

First, genetic variance (V_G) can itself be further partitioned into additive variance (V_A), dominance variance (V_D), epistatic variance (V_{EP}), and variance due to assortative mating (V_{AM}). V_A includes independent genetic effects that get fully transmitted from parents to offspring. V_D includes phenotypic differences arising from interactions of different alleles at the same locus, while V_{EP} subsumes interactions of genes at different loci. V_{AM} measures the degree to which variance is increased or decreased through non-random pairing of parents' genotypes. Total genetic variance can then be expressed as the sum:

$$V_G = V_A + V_D + V_{EP} + V_{AM}$$

The ratio of total genetic variance to phenotypic variance (V_G/V_P) is called "heritability in the broad sense" or "coefficient of genetic determination."

---

[7] Table 1.1 is a way to explain the *meaning* of "heritability," but it does not correspond to the way heritability is calculated in empirical practice. Usually we do not know the number of genotypes, let alone the numerical values of genotypic means, and the same is true of the environmental side. Still, this idealization is useful for getting some basic conceptual understanding, and it also prepares the ground for discussing issues about genotype–environment interaction (in chapter 2).

On the other hand, the so-called "heritability in the narrow sense" is the ratio of *additive* genetic variance to phenotypic variance (V_A/V_P). Since in this book I will primarily address issues in which broad heritability is more relevant, from now onwards the term "heritability" will be understood as *broad* heritability (unless specified otherwise).

Second, additional complications in the variance story arise from other possible sources of variation (beside V_G and V_E). These include gene–environment interaction or V_I (to be discussed in chapter 2), gene–environment correlation or Cov_{GE} (to be discussed in chapter 3), and measurement error or V_{Error}. Finally, in contemporary behavior genetics, environmental variance (V_E) is usually also divided into two parts: between-family variance (V_{BF}) and within-family variance (V_{WF}). Between-family variance and within-family variance are sometimes referred to, respectively, as being due to shared (or common) and non-shared (or unique) environments. Summing all this up, total phenotypic variance can be represented in the following general way:

$$V_P = V_A + V_D + V_{EP} + V_{AM} + V_{BF} + V_{WF} + V_I + Cov_{GE} + V_{Error}$$

Of the two ways of understanding heritability, narrow and broad, I said I will focus on the latter. But I have to mention the *third* interpretation of heritability, which also exists in the literature and which sometimes creates confusion. Some biologists and philosophers (Lehrman 1970: 22; Roughgarden 1979: 136; Lerner 1986: 129; Sober 1984: 151; Feldman 1992: 151; Brandon 1996: 70, 93) regard a trait as heritable if there is a requisite kind of resemblance between parents and offspring – irrespective of whether that resemblance is due to genetic or environmental causes. There is nothing wrong with that definition, of course. As Jacquard (1983) says, it just happens that there is one word ("heritability") but three different concepts: (1) broad heritability (the ratio of genetic variance to total variance), (2) narrow heritability (the ratio of *additive* genetic variance to total variance), and (3) heritability in the sense of parents–offspring resemblance.

In terms of ordinary language semantics, (3) comes probably closest to what the word "heritable" means in English. Moreover, under certain assumptions, (3) can be regarded as an "operational definition" of (2). That is, when it is hard to measure narrow heritability directly, under some conditions one can measure (3), the regression of offspring phenotype on mid-parent phenotype, and take it as an estimate of (2), narrow heritability. Sense (1), on the other hand, has much less to do with the ordinary meaning, and it is therefore a rather unfortunate historical accident that the term "heritability" was chosen to designate that concept. Hence, it cannot be stressed enough that the reader should always think about the notion by relying on its definition, and not on what it means in English.

The troubles start when people switch from one concept to its homonym without noticing it. Here is an example from one of the widely used introductions to philosophy of biology (the numbering of the two sections is added for easy contrast):

> [1] It is a common mistake to think that high heritability means that a trait is genetically determined – a matter of nature rather than nurture. In fact, heritability has very little to do with how traits are built in the growing organisms. Selection cares about whether your children resemble you. But it doesn't care why. Heritability is purely a measure of how well the state of the parent predicts the state of the offspring ... [2] One way to make a trait highly heritable is to make the environment the same for everyone. By controlling other causes of variation, we can make heritable variation a higher proportion of total variation. For example, IQ scores will be more heritable if we provide equality of educational opportunities. Conversely, genetic uniformity in a population will reduce heritability. (Sterelny & Griffiths 1999: 35)

Sterelny and Griffiths are here criticizing the muddled thinking about heritability (the box from which the quotation comes is entitled "A Caution on Heritability"), but in the end they manage to contradict themselves in the space of a single page. First, they insist that heritability is *neutral between genetic and environmental influences* (because it is "purely a measure of how well the state of the parent predicts the state of the offspring"), and then after just a few sentences they say that *the existence of genetically caused differences increases heritability*. The source of this contradiction is fairly clear. In section [1], Sterelny and Griffiths have in mind the third concept of heritability (parent–offspring resemblance), and in section [2] they speak about variance-defined heritability. But it is simply a confusion to run these two concepts together.

The heritability claims have been regarded as important primarily because it was expected that they would furnish some valuable information about the *causal* strength of genetic influence on phenotypic differences. Let us recall that J. L. Lush, who is credited (wrongly) with introducing the term "heritability," defined it "as the fraction of the observed variance which was *caused* by differences in heredity" (quoted in DeFries 1967: 324 – emphasis added). Sewall Wright, who introduced the symbol h^2 for heritability in 1920, defined it as "the degree of determination by heredity" (quoted in Bell 1977: 297).

There is a big methodological divide between two domains of heritability research: studies of animals and humans. Animal studies use the experimental approach. Here organisms with different genotypes can be exposed to a number of different environments, and then observing the phenotypes resulting from all G–E combinations makes it relatively easy to infer the causal impact of the two factors. In human studies such an approach is impossible for ethical reasons. Behavior geneticists cannot assign babies with different genetic characteristics to different environments just to satisfy their scientific curiosity. They had to find another way to do it. And they did. Twin studies, studies of relatives of different degrees of genetic affinity, and adoption studies are the best known methods of human behavior genetics. For a detailed exposition of more advanced methods see Neale & Cardon 1992. As David Fulker (1974: 91) said, it was precisely the absence of experimental control and the complexity of human behavior genetics that forced the field to develop models of high statistical sophistication. (For a more accessible overview of contemporary methods see Plomin et al. 2001: 327–371, and also very nice "behavioral genetic interactive modules," developed by Shaun Purcell and freely available at http://statgen.iop.kcl.ac.uk/bgim/.)

This fundamental methodological difference between animal and human studies was a point where many critics tried to drive a wedge between the two approaches. They claimed that the search for heritability makes sense in the animal domain because there the experimental manipulation of the two variables makes it possible to separate their causal contributions. But in research on humans, they insisted, there is an inherent methodological limitation that dooms the whole enterprise in advance. For example, Oscar Kempthorne said: "The *only* way to infer causality, and then only perhaps a very limited sort of causality, is by way of controlled experiments, which are essentially impossible in the human genetic-environment milieu" (in Pollak et al. 1977: 13 – italics added; cf. Kempthorne 1997: 111; Platt & Bach 1997: 139).

It is in the context of human behavior genetics that it has been said that the term "heritability" is "no longer suitable for use in *human genetics* and its use should be discontinued" (Guo 2000: 299); that to apply the heritability formula to humans is "virtually impossible" (Park 2002: 407); that heritability estimates are "both deceptive and trivial" (Hirsch 1976: 168); that they are "nearly equivalent to no information at all for any serious problem of human genetics" (Feldman & Lewontin 1975: 1168); that they are "unscientific and, indeed, meaningless" (Layzer 1976: 199); that they have "as much scientific validity as horoscopes" (Layzer 1976: 239); that it is dubious whether a clear meaning can be given to "genetic determination of traits" (Burian 1981: 51); that inferences about genetic determination of traits should be "disavowed once and for all" (Kitcher 1990: 97); that "mathematical estimates of heritability tell us almost nothing about anything important" (Jencks et al. 1972: 76); that the attempt quantitatively to determine the part of the phenotypic variance due to genetic causes is "biological nonsense" (Lewontin 1982: 14–15); that the concept of heritability is "questionable" (Schaffner 1999: 61), or "an imposing illusion" (Moore 2001: 40); that "the next century will treat heritability analysis with the same regard that this one treats phrenology" (Sarkar 1999: 230); that heritability analysis "ought to be relegated to the history of science along with phlogiston, penis envy and cold fusion" (Wahlsten 1994: 265); that "[the 'facts' from which the heritability of IQ has been calculated] have been so scandalously bad as to constitute a veritable Watergate of human behavioral genetics" (Lewontin 1976b: 10); that the talk about substantial heritability of IQ is *scientifically meaningless garbage*" (Lewontin 1973 – italics in the original), and so forth.

Two central arguments drive this methodological pessimism: one based on the possibility of genotype–environment interaction, and another relying on the possibility of genotype–environment correlation. These two objections will be extensively discussed in chapter 2 and chapter 3, respectively.

Another reason for many people's skepticism about heritability is clearly articulated by Patrick Bateson: "Another problem with heritability is that it says nothing about the ways in which genes and environment contribute to the biological and psychological cooking processes of development" (Bateson 2001b: 565). Indeed, heritability is supposed to give us causal knowledge, but it gives us no information about causal details or mechanisms underlying the process of development. But it is wrong to criticize a concept for not delivering something that it was not supposed to deliver. The very purpose of heritability estimates is to give us some knowledge about causality precisely in those situations where we are ignorant of causal specifics.

Arthur Jensen once said that "a heritability study may be regarded as a Geiger counter with which one scans the territory in order to find the spot one can most profitably begin to dig for ore" (Jensen 1972b: 243). That Jensen's advice as to how to look upon heritability is merely an application of a standard general procedure in causal reasoning is confirmed by the following observation from an introduction to causal analysis: "the decomposition of statistical associations represents a first step. The results indicate which effects are important and which may be safely ignored, that is, where we ought to start digging in order to uncover the nature of the causal mechanisms producing association between our variables" (Hellevik 1984: 149). High heritability of a trait (in a given population) often signals that it may be worthwhile to dig further, in the sense that an important genetic mechanism controlling differences in this trait may thus be uncovered.[8]

The attraction of the heritability approach is that it is supposed to give us some causal knowledge even though we are unable to peek inside the black box and see how the cogs and wheels fit together. And if it can do what it is supposed to do, is it not odd to complain that it did not do more? We usually do not see it as the fault of the hammer that it cannot smash the atom.

Several critics of heritability (Hirsch 1997: 208, 213; Hirsch 1990: 137; Wahlsten 1997: 73; Sarkar 1998: 72; Schiff & Lewontin 1986: 179; Daniels et al. 1997: 53; Gigerenzer 1997: 145; Stoltenberg 1997: 90; Guo 1999: 227; Guo 2000: 298; Vitzthum 2003: 544) use an argument from authority to prop up their position. They all quote Ronald A. Fisher, who once called heritability "one of those unfortunate shortcuts, which have often emerged in biometry for lack of a more thorough analysis of the data" (Fisher 1951: 217). There are five problems here.

First, all of these authors invoke Fisher in the context of their discussion of human behavior genetics, whereas Fisher's paper addressed issues connected with animal breeding problems. This is odd because most of these critics insist that it is *human* behavior genetics that suffers from specially grave methodological defects, while they are usually happy to concede that heritability works fine in experimental studies of animals (like cattle breeding, for example). So, Fisher's argument seems to be an inappropriate instrument for their purpose.

Second, Fisher certainly did not argue that heritability is useless and that the concept should be abandoned by geneticists. After all, he himself spoke about heritable variation (Bennett 1983: 140), about the genetic component of variance vs. total phenotypic variance (Bennett 1983: 228), etc. Even in his historical book *The Genetical Theory of Natural Selection*, Fisher speaks about "the fraction of the total observable variance of the measurement, which may be regarded as genetic variance" (Fisher 1930: 33), which is nothing if not heritability by another name. (He left that part

---

[8] "Discovering [components of variance] is a necessary first step, but it is just a first step, and few would argue that it is sufficient in itself" (McGuffin & Katz 1990: 142).

unchanged in the significantly revised second edition of 1958.) Elsewhere, again, Fisher praised Galton for trying to give numerical precision to concepts like "the strength of inheritance" (Fisher 1959: 2), and as we know, "Galton's original question about the 'strength' of inheritance is the same as the contemporary common sense understanding of heritability" (Nuffield 2002: 20).

Third, Fisher's disparaging comment about heritability was far from being universally accepted. His worry was mainly about the heritability value being under undue influences of measurement error and changes in population parameters. For instance, his view was criticized by Johansson for "missing the important point that for practical breeding work we need estimates of heritability that are valid under the conditions where they will be applied" (Johansson 1961: 9).

Fourth, Fisher's grounds for dissatisfaction with heritability were completely different from the reasons of the critics who quote him in support of their own views. In the opinion of most of the authors who cite Fisher, one of the main problems with heritability is that genetic variance (the numerator) cannot be determined because of the confounding influence of gene–environment interaction. But Fisher himself had nothing but contempt for this kind of criticism, as seen from his letter to the son of Charles Darwin in 1935: "There is one point in which Hogben and his associates are riding for a fall, and that is in making a great song about the possible, but unproved, importance of non-linear interactions between hereditary and environmental factors. J. B. S. Haldane seems tempted to join in this" (Bennett 1983: 260). It is rather funny that contemporary critics of heritability (some of them doing research in history of biology!) press the interaction objection, which Lewontin basically took over from Hogben (1933), and then even present Fisher as their ally in this. By the way, Fisher's judgment that Haldane "seems tempted to join in this" proved correct. Haldane's book in which he defended Hogben's position on interactions soon followed (Haldane 1938).[9]

Fifth, while there is nothing wrong with these authors trying to support their opposition to human behavior genetics by relying on Fisher as a supreme authority in genetics (after all, he was one of the architects of

---

[9] Similarly, when Medawar says that objections to heritability "seem to be beyond the comprehension of IQ psychologists, though they were made clear enough by J. B. S. Haldane and Lancelot Hogben" (Medawar 1977a), he did not consider the possibility that IQ psychologists were intelligent enough to understand the objections but that, like Fisher, they were simply not convinced by them.

the Modern Synthesis), they should be aware that they might not find his other views about genetics so appealing. Some of them would probably be shocked to know that, despite his derogatory aside about heritability, Fisher actually believed that genetics supports the heritability of racial differences in intelligence. At about the same time that he wrote the article they quote, he suggested the following formulation for the UNESCO statement on race: "Available scientific knowledge provides a firm basis for believing that *the groups of mankind differ in their innate capacity for intellectual and emotional development*, seeing that such groups differ undoubtedly in a very large number of their genes" (quoted in Provine 1986: 875 – italics added). This shows that one should be careful in choosing one's authorities because they can force one to go in an unwanted direction.

It is worth stressing that influences of genes and environment are inseparable epistemologically: we cannot know one without the other. For this reason it is wrong to regard the interest in heritability as a symptom of an obsession with genes. By knowing heritability one knows *ipso facto* the possible size of complementary contribution of environment to phenotypic variation, which is called "coefficient of environmental determination" (DeFries 1972: 12) or "environmentality" (Plomin et al. 2001: 297).

It is hence curious to see heritability attacked by people who want to emphasize the importance of environmental influences. For, if their criticism succeeds in making the heritability methodologically suspect or scientifically useless, they are thereby pulling the rug from under their own feet: they have then no right to speak about the environmental impact on phenotypic variation, either. A concept complementary to a meaningless concept is itself meaningless.

Some scholars adamantly reject the possibility of separating the causal contributions of heredity and environment, but without noticing any logical inconsistency they at the same time cheerfully advocate environmentalist explanations. For example, in an article written for the APA *Encyclopedia of Psychology*, Douglas Wahlsten (2000b: 382) criticizes methods of behavior genetics and claims that its partitioning of phenotypic variance into genetic and environmental components is "flawed in principle." Then in the very next sentence he says that behavior genetic studies "have valuable applications in the study of nongenetic effects." Apparently, the partitioning of variance is flawed only in one direction: nature is "in principle" inseparable from nurture, whereas somehow nurture turns out to be quite easily separable from nature.

1.4 CAN MONOMORPHIC TRAITS BE HERITABLE?

A frequent objection to heritability is that although it is supposed to be a measure of genetic influence, some traits have zero heritability despite being obviously under genetic control. These are monomorphic traits, i.e., those possessed by all members of a given population.

Take the example of the human characteristic of walking on two legs: the trait seems to be under genetic control, while nearly all the variation in this trait is presumably the result of environmental disturbances. The trait is genetic but not heritable. Some people (Ariew 1996: S23; Bateson 2001a: 151–152; Bateson 2001b: 565; Dupré 2001: 30; Dupré 2003: 105; Vitzthum 2003: 544) think that this consequence is a serious problem for heritability, mainly because it shows that even despite zero heritability genes can still have an important role to play in development. But this is a serious misunderstanding for two reasons.

First, no geneticist or behavior geneticist ever took zero heritability to mean that genes are unimportant for development. What *is* taken to follow from zero heritability is only that genes are unimportant for explaining the existing phenotypic *differences* in the population in question.

Second, even if a trait is shared by *all* organisms in a given population it can still be heritable – if we take a broader perspective, and compare that population with *other* populations. The critics of heritability are often confused, and switch from one perspective to another without noticing it. Consider the following "problem" for heritability:

> the heritability of "walking on two legs" is zero. And yet walking on two legs is clearly a fundamental property of being human, and is one of the more obvious biological differences between humans and other great apes such as chimpanzees or gorillas. It obviously depends heavily on genes, despite having a heritability of zero. (Bateson 2001b: 565; cf. Bateson 2001a: 150–151; 2002: 2212)

When Bateson speaks about the differences between humans and other great apes, the heritability of walking on two legs in *that* population (consisting of humans, chimpanzees, and gorillas) is certainly not zero. On the other hand, within the human species itself the heritability may well be zero. So, if it is just made entirely clear which population is being discussed, no puzzling element remains. In the narrower population (humans), the question "Do genetic differences explain why some people walk on two legs and some don't?" has a negative answer because there are no such genetic differences. In the broader population (humans,

28

chimpanzees, and gorillas) the question "Do genetic differences explain why some organisms walk on two legs and some don't?" has an affirmative answer. All this neatly accords with the logic of heritability, and creates no problem whatsoever. The critics of hereditarianism like to repeat that heritability is a population-relative statistic, but when they raise this kind of objection it seems that they themselves forget this important truth.

The same pseudo-problem for heritability is created by Susan Oyama. After describing a standard situation where gene differences explain phenotypic differences, she tries to show that sometimes genes are in an unprincipled way also invoked to explain *lack of differences*. Her example is "a species universal, like the human smile, or walking" (Oyama 2000b: S334). But again, in explaining species universals, what we account for is why a particular species has a given trait, while other species don't. It is a question about (inter-species) variance, not *in*variance. And the genetic answer, far from violating "causal democracy" (the general symmetry between the roles of genes and environments), is entirely in accordance with it.

André Ariew falls into the same trap. He says that "on the proposal that innateness is high heritability, the possession of opposable thumbs [which has low heritability in human population] is in this case (counterintuitively) not innate" (Ariew 1996: S23). In fact, there is no counterintuitive result at all. The low heritability of opposable thumbs among humans reflects the fact that, indeed, in *this* population (humans) there is no genetic variation in that trait. However, the reason we regard the trait as genetic is that we actually think about the variation in this trait in a wider population (say, consisting of humans and monkeys like tamarins and marmosets). In *that* population, heritability of having opposable thumbs is high, and the trait is genetic.[10]

Let me try to make the point in a very general way, by using a highly simplified example of a population of organisms with just two genotypes (G_1 and G_2) living in two different environments (E_1 and E_2). Suppose that they all have the same phenotypic trait, P_1, which means that there is no phenotypic variation with respect to that trait. Now if someone nevertheless says that in that population trait P_1 is genetic, what could that possibly mean? Apparently, it cannot mean that the trait is heritable because the heritability of P_1 is not well defined in this case. Since h^2 is

---

[10] It is odd that Ariew did not consider this approach, given that Fred Gifford (1990: 336) defended it using the very same example.

the ratio of genetic variance to the entire variance, and since both values are zero here, the talk about heritability seems to make no sense at all.

But let's not move too quickly here. If the claim that the universally shared trait P_1 is genetically caused means anything, it means that it is the actual genetic constitution of these organisms that explains why they all have P_1, instead of having some other trait (say, P_2) or a distribution of other traits (P_i, P_j, P_k ...). In other words: *had the genetic constitution of the population been different*, not all the organisms would have had P_1. So, in attributing the monomorphic trait to genes we are actually saying that *different* genes would have resulted in a *different* phenotypic structure of the population. We are in fact contrasting the existing population with another population (actual or possible, depending on the context) in which different genes produce a different phenotypic outcome. Without such a contrasting case, I submit, it is difficult to make sense of the statement that P_1 is genetically caused.

If this is correct, then it will turn out that the talk about genetic causation with respect to monomorphic traits is intelligible only if there is heritability – in a wider context. Namely, what happens when we contrast the existing population with another one is that our perspective is broadened and, most importantly, something is introduced that has not existed earlier: *variation*. In the expanded situation, which includes both the original population and its contrast case, it becomes entirely legitimate to look for the source of the phenotypic *difference* between them. And if the source is genetic, it will be because the difference in question *is* heritable.

Figures 1.2a and 1.2b show one and the same population (solid line) contrasted with two different possible situations (dashed lines). The answer to the causal question about what produces P_1 is different in the two figures because the contrast cases are different. In Figure 1.2a, P_1 is genetic simply because it is the variation in genotype, (G_1 or G_2) vs. G_3, that determines whether an organism will have P_1 or P_2. In Figure 1.2b, on the other hand, P_1 is environmental because organisms have this phenotype due to being exposed to environments E_1 or E_2, rather than E_3 (which leads to P_2).

Figure 1.2a basically corresponds to Bateson's example (mentioned above), where walking on two legs is regarded as genetic because the *genetic difference* between us and chimpanzees explains why we have the trait and chimpanzees do not. The variation in the way of locomotion is heritable in the population that comprises humans *and* chimpanzees. Notice, however, that if we stick to the example of "walking on two legs"

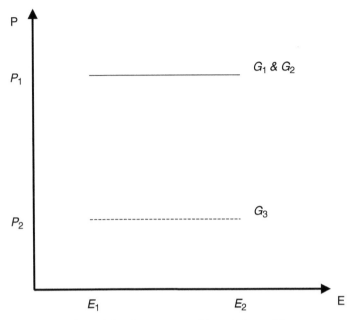

Figure 1.2a A monomorphic trait (genetic).

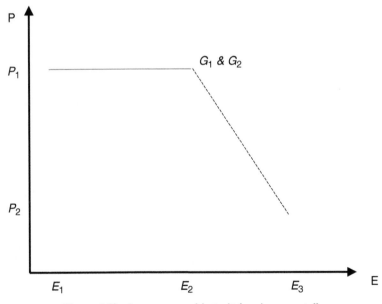

Figure 1.2b A monomorphic trait (environmental).

there is no readily available contrast case that would correspond to the situation represented by Figure 1.2b. Here we have to resort to our imagination and picture a population in which, say, all people contract polio (environmental influence) and become paraplegics. Compared to *that* situation, our walking on two legs would indeed be environmental.

So, the puzzle about how a monomorphic trait (for which heritability is undefined) can still be regarded as genetic is resolved by realizing that in such cases we always expand the population (be it only implicitly), thereby creating a phenotypic variation which did not exist in the first place and which can then be explained by genetic differences. In other words, a monomorphic trait can be regarded as genetic simply because in the broadened context it is no longer monomorphic and is *heritable*. My claim is that if we look at monomorphic traits *without* introducing a contrast case (and variation that goes with it), the question "Genetically or environmentally caused?" cannot receive a meaningful answer.[11]

Consider the same solid line in Figures 1.2a and 1.2b and forget completely about contrast cases pictured by two alternative dashed lines in these figures. Now focusing *exclusively* on what that solid line represents – the population of four types of organisms ($G_1 + E_1$, $G_1 + E_2$, $G_2 + E_1$, $G_2 + E_2$), which *all* have the same phenotype P_1 – try to ask yourself whether trait P_1 is genetic or environmental. What could possibly sway you one way or the other? Without recourse to additional information (a contrast case) the question is unanswerable.

This diagnosis can help us avoid a pseudo-problem for heritability created by Ned Block. He argues that the concept *heritable* is different from the commonsense concept *genetically determined*. He first points out that we would all agree that the number of fingers on the human hand is *genetically determined*, but then he finds it puzzling that the *heritability* of the number of fingers in humans is very low, and exclaims: "What's going on?" (Block 1995: 103). Well, here's what's going on. Block uses the concept "genetically determined" in such a way that he *explicitly* links it to the *normal* environment ("it depends on the idea of a normal environment"). But then, of course, if the environment is fixed to a normal environment (which never produces an abnormal number of fingers), it follows that the only possible contrast case (*not* having five fingers on each hand) can be the organisms that are phenotypically different because they are *genetically different*. And clearly, once there is a genetically produced

---

[11] Essentially the same point, I think, is made in a different way by Elliott Sober in Sober 1994: 194–195.

phenotypic difference, heritability reappears. Contrary to what Block says, the trait being genetically determined turns out to be inseparable from its being heritable.

There are two sources of Block's confusion here. First, he does not realize that if the concept of a *genetically determined* trait P is semantically linked to what happens in a *normal environment* (which is not a source of variation for the trait in question) then the only remaining possible reason for an organism *not* to have P must be a genetic difference – which therefore makes it certain that any phenotypic difference emerging under the stipulated conditions will be heritable. Second, Block fails to see that, in this context, the postulation of a genetic difference is essential for attributing any causal role to genes ("genetically determined"). For instance, when in a later footnote (1995: 109) he claims that there is no guarantee that it will always be possible to expand the relevant population along the lines that correspond to my Figure 1.2a, he may be right about this. But without such an expansion that would point to a genetic difference, there would be absolutely no reason to say that trait P is genetically, rather than environmentally, determined.

To drive this point home, consider a population in which everyone suffers from a given disease, and let us also assume that in this population everyone has the same genotype G_1 and is exposed to the same environment E_1. Now is this disease genetic or environmental? The truth is that the question just makes no sense unless we are allowed to go beyond that situation and explore whether the disease would still be there under a *different* regime, say, with some members of the population having genotype G_2 (instead of G_1), or environment E_2 (instead of E_1). No difference, no comparison, no causal attribution.

1.5 PHILOSOPHERS AT WORK: *CAVEAT LECTOR!*

In those scientific controversies where methodological issues loom large, philosophers rightly feel that their contribution might be helpful. Their careful analysis of opposing views promises to unearth some unrecognized points of disagreement, help the sides in the conflict to understand each other's positions better, and eventually maybe remove some obstacles to a more fruitful exchange of ideas. It is all the more disappointing, therefore, that here where philosophical intervention is really needed it has so often just compounded the confusion. George Berkeley's famous diagnosis of philosophers' troubles applies here very well:

> Upon the whole, I am inclined to think that the far greater part, if not all, of those difficulties which have hitherto amused philosophers, and blocked up the way to knowledge, are entirely owing to ourselves – that we have first raised a dust and then complain we cannot see. (Berkeley 1996: 8)

Let me illustrate how the anti-hereditarian consensus in philosophy (mentioned in the Introduction) led the field to be completely dominated by highly prejudiced opinions, uninformed statements, and occasional downright ignorance. It is almost as if there was a competition in pooh-poohing heritability research, in which everyone won a prize, just for participating.

Many philosophers represent the nature–nurture debate as being the result of dreadfully crude thinking, or even occasionally as "having little to do with thinking" (Blackburn 2002). In the attempt to reduce to absurdity the notion that the relative importance of inheritance and environment could be measured, Gilbert Ryle observed that, in speaking about Edmund Hillary's climbing Everest, we are not permitted to say "that Hillary's inherited physique got him up 15,000 feet, leaving the remaining 14,000 feet to be contributed by his acquired expertness" (Ryle 1974: 57). Of course, it would be silly to say this, but it is curious that the evident silliness of that claim did not lead Ryle to draw the only logical inference, i.e., that serious people who tried to gauge causal contributions of nature and nurture must have had in mind something different from the ridiculous idea he "refuted."

1.5.1 *g or not g, that is the question*

The heritability of IQ is the best researched area in the heredity–environment controversy (McGue & Bouchard 1998: 4; Plomin 1987: 51; Herrnstein 1973: 53; Plomin 2002: 212). It is therefore interesting to take a look at some typical philosophical excursions into the topic of human intelligence, and especially "general intelligence," which some influential writers (see 1.5.2) closely associate with discussions about heritability. Gilbert Ryle disputed the belief in general intelligence (Spearman's *g* factor) in the following way: "Only occasionally is there even a weak inference from a person's possession of a high degree of one species of intelligence to his possession of a high degree of another" (Ryle 1974: 55). This statement is empirically false. In psychology, there has been a consensus for quite some time that different cognitive abilities *are* substantially correlated. Nevertheless, we should not make too much of Ryle's mistake. It is very uncharacteristic of his way of doing philosophy to issue statements about empirical issues. It goes against his view that philosophy should be insulated from empirical considerations and be limited to conceptual analysis or, as he used to put it, "logical geography." What most probably happened is that Ryle (wrongly) regarded the independence of different forms of intelligence as an obvious truth, and then just used this alleged commonsense truth to make a "conceptual" point against the idea of general intelligence.

More worrying, however, is when the same mistake is repeated by contemporary philosophers who are trained to pay attention to results of the relevant empirical research. Philip Kitcher, relying on a paleontologist and two philosophers as his sole authorities on the research on human cognitive abilities, proclaimed general intelligence a "myth" (Kitcher 1985: 200–201). To see how far off-base his judgment is, notice that the American Psychological Association's task group on intelligence stated that the theory of general intelligence is "the most widely accepted current view" (Neisser et al. 1996: 81). Also, g is regarded as "perhaps the most replicated result in psychology" (Deary 2000: 318), and as "probably the best measured and most studied human trait in all of psychology" (Gottfredson 2002: 25). Another prominent researcher on psychology of intelligence calls the general intelligence factor "certainly the most robust phenomenon in the social sciences," and adds: "Despite torturous method of factor analysis, attacks from outraged critics and even long periods of being ignored, g [general intelligence] just keeps reappearing like the insistent relative that won't go away" (Detterman 2000: 136). In a recent conference attended by leading experts on human intelligence Michael Rutter, who is known for his moderate views, said: "*All of us* accept [g's] reality. It is not merely a statistical artefact, rather, it really does represent something that is biologically important" (Rutter 2000: 282 – italics added).

But it is not merely that Kitcher's view on g is out of touch with contemporary science. His reason for dismissing g is extremely puzzling: "Many scientists are now convinced that there is no single measure of intellectual ability – no unitary intelligence. Their suspicion of the concept of intelligence is based on the view that various intellectual capacities are not well correlated" (Kitcher 1985: 200–201). Kitcher is simply uninformed here. For example, the correlations between four different intellectual abilities measured by WAIS III (Wechsler Adult Intelligence Scale III) are between 0.6 and 0.8. In social science one rarely gets correlations higher than that. The correlation between various cognitive abilities has been such a consistent and robust result of measurement since the 1940s that today even the psychologists who oppose g never deny this fact. For example, one of the most prominent *critics* of general intelligence had to concede: "The evidence in favor of a general factor of intelligence is, in one sense, overwhelming ... One would have to be blind or intransigent not to give this evidence its due" (Sternberg 2003: 375).

Hilary Putnam also asserted, without supplying any references, that all the statistical evidence is against the hypothesis of a single factor (1973: 141). John Dupré even goes so far as to suspect that no coherent sense can be attached to the concept of general intelligence (Dupré 2004: 513). He also claims that any measurement that ranks individuals on a single scale of intelligence must include some weighting of specific cognitive abilities, and that this weighting "can surely only be a value judgment." But this is simply false. In fact, the scale of general intelligence is created by taking an IQ correlation matrix and then extracting g from that matrix as the first principal component (that accounts for the highest proportion of IQ variance). This statistical procedure is usually done by a computer, without the machine being fed any "value judgment" about specific cognitive abilities.

Again, a statement that very much sounds like a summary (and uninformed) rejection of general intelligence has found its way even into the *Oxford Dictionary of Philosophy* (see Blackburn 1996: entry "intelligence"). Also, despite having no obvious expertise in the relevant research areas, a dozen philosophers from Ohio State University published a short letter in 1990, in which they claimed that, "as is now well known," the g factor accounts for little variance in IQ (Boer et al. 1990). Judging by the similarity of wording and reliance on L. L. Thurstone as an authority, it appears that the philosophers' pronouncement derived from Stephen Jay Gould's claim to the same effect (Gould 1981: 314–315). In reality, however, the g factor typically explains about half of the variation in IQ (Carroll 1995; Deary 2001: 222; Deary 1998; Lubinski 2004: 98), quite a large effect by any measure. Interestingly, what almost a whole department of philosophy publicly declared to be "now well known" is regarded by the world's leading expert on factor analysis as "truly an egregious error on Gould's part" (Carroll 1995). Even the prominent *opponents* of g, like Ulric Neisser, reject Gould's views because they regard his critique of general intelligence "as actually rather thin, relying chiefly on rhetoric and ignoring empirical evidence" (Neisser's response to a question in the PBS Newshour Online Forum, on April 28, 1998).

Why do the philosophers present the research on *g* and its heritability as a failure and an irrational enterprise, despite plenty of evidence that it is in reality a bustling and fruitful research program? If we accept Paul Meehl's diagnosis of why the very idea of general intelligence still faces obstinate resistance, there are three possible explanations of philosophical *g*-phobia: "A century of research – more than that if we start with Galton – has resulted in a triumph of scientific psychology, the foot-draggers being either uninformed, deficient in quantitative reasoning, or impaired by political correctness" (Meehl 1998).

1.5.2 Measure for (mis)measure

In philosophy of science, Gould's book *The Mismeasure of Man* is standardly praised as disclosing serious weaknesses and fallacies of hereditarianism. Here are just a couple of representative examples:

> No one has done as much as Stephen J. Gould to expose race and intelligence studies for the garbage that they often are. (Brown 1998: 5; 2001: 206)

> Stephen Jay Gould has lucidly analyzed how filling the skulls with lead shot, and comparing the weights of the lead, could easily be infected with unconscious biases. (Kitcher 1996: 171)

Brown is urging others to follow in Gould's footsteps and fight bad science in the same way. He says that philosophers of science are "uniquely situated" to do this job well: "More than anyone else, they have the skills – logical, mathematical, statistical, methodological, and many more – to ferret out bad science" (Brown 1998: 5).

Perhaps. But are the views attacked by Gould really bad science? Brown, Kitcher, and many others have no doubt about this although they provide no supporting evidence for this belief other than Gould's authority. However, *The Mismeasure of Man* is quite controversial as a piece of scholarship. The reviews of Gould's book in *Nature*, *Science*, and some other professional journals were highly negative and severely critical (see Davis 1986), in contrast to typically favorable and laudatory comments in the popular press. Therefore, although it still remains perfectly legitimate for philosophers of science to side with Gould and express their admiration publicly, one would expect them, if nothing more, then at the very least to notify the reader about the massive presence of these strong dissenting voices as well (assuming of course that they are aware of them).

For illustration, here are some extracts from the review in *Nature* written by Steve Blinkhorn:

> With a glittering prose style and as honestly held a set of prejudices as you could hope to meet in a day's crusading, S. J. Gould presents his attempt at identifying the fatal flaw in the theory and measurement of intelligence. Of course, everyone knows there must be a fatal flaw, but so far reports of its discovery have been consistently premature . . . [*The Mismeasure of Man*] is a book which exemplifies its own thesis [that science is necessarily influenced by social prejudices]. It is a masterpiece of propaganda, researched in the service of a point of view rather than written from a fund of knowledge . . . But verbal fluency is no substitute for good arguments in the long run. The substantive discussion of the theory of intelligence stops at the stage it was in more than a quarter of a century ago . . . Gould even gives a perfectly straightforward account of what heritability would and would not mean in terms of the modifiability of intelligence, but fails to point out that such arch-hereditarians as Eysenck and Jensen have published essentially identical accounts . . . The truth of the matter is that Gould has nothing to say which is both accurate and at issue when it comes to substantive and methodological points . . . [Many of his assertions] have the routine flavor of Radio Moscow news broadcasts when there really is no crisis to shout about. You have to admire the skill in presentation, but what a waste of talent. (Blinkhorn 1982: 506)

More to the point, however, Gould's central argument against hereditarians happens to be based on a misunderstanding of the position he is criticizing. He says: "a reified Spearman's g is still the only promising justification for hereditarian theories of mean differences in IQ among human groups . . . The chimerical nature of g is the rotten core of Jensen's edifice, and of the entire hereditarian school" (Gould 1981: 320). In reality, Jensen's views on the genetic explanation of racial differences in IQ are *totally independent* from the question whether there is only one factor of intelligence or more factors. Here is what James Flynn, a consistent critic of Jensen, had to say on the matter:

> Gould's book evades all of Jensen's best arguments for a genetic component in the black-white IQ gap, by positing that they are dependent on the concept of g as a general intelligence factor. Therefore, Gould believes that if he can discredit g, no more need be said. This is manifestly false. Jensen's arguments would bite no matter whether blacks suffered from a score deficit on one or 10 or 100 factors. I attribute no intent or motive to Gould, it is just that you cannot rebut arguments if you do not acknowledge and address them. (Flynn 1999a: 373)

Another argument against hereditarianism in *The Mismeasure of Man* has no problems with misinterpretation or logic. However, it seems to have serious troubles with empirical reality (although the news about this is traveling slowly, and has not yet reached mainstream philosophy of science). The argument that made Gould's book famous and that left the strongest impression on many readers is certainly his criticism of the skull measurements undertaken by the nineteenth-century scientist Samuel George Morton. Gould claimed that the results of Morton's measurements indicating systematic differences in cranial capacity between different races were due to Morton's unconscious bias, and ultimately his racist beliefs: "Morton's summaries are a patchwork of fudging and finagling in the clear interest of controlling a priori convictions" (Gould 1981: 54). Gould then went on to propose a concrete explanation of how the bias worked to distort the measurements (e.g., Morton's pushing mightily with his thumb in the attempt to fill Caucasian crania with more seed, and make them appear larger). It is this account that Kitcher called "lucid."

Now, elementary logic demands that if you want to argue that someone's mistake is due to some kind of bias or prejudice, you have first to be sure that the person really made a *mistake*. In the case of Morton's measurements there appears to be no room for doubt about his having made the mistake. For, the idea that human races differ in average cranial capacity or brain size sounds to many people like the crudest possible form of racist and pseudo-scientific belief. But notice that the belief is nevertheless empirical, and that its truth-value cannot be determined by conceptual analysis or political condemnation. John S. Michael thought that it was worth checking the data, and in 1986 he remeasured the cranial capacities of 201 specimens from the Morton Collection. In a paper published in *Current Anthropology* (Michael 1988) he presented the results, and showed that the differences reported by Morton were basically *corroborated* by his remeasurements. Although Michael had qualms of a more general kind (e.g., about the legitimacy of "race" as a biological category), with respect to the issue at hand (the craniological data) his conclusion was that he could find no indication of the systematic bias Gould ascribed to Morton, and that in his opinion "Morton's research was conducted with integrity" (Michael 1988: 353).

Gould's explanation of Morton's "error" in terms of racial bias fails for the simple reason that there is no error that needs to be explained (i.e., there is no *explanandum*). In other words, what Morton discovered was a genuine difference. Moreover, this fact is accepted today in standard reference books (see Sternberg 1982: 773; Brody 1992: 301; Mackintosh

1998: 184), and even by scholars who are staunch advocates of the environmentalist account of the racial IQ gap. For instance, Ulric Neisser (one of the leading critics of the genetic hypothesis and the chair of the American Psychological Association Task Force that prepared the report "Intelligence: Knowns and Unknowns") did not hide his strong aversion to the hereditarian views of J. Philippe Rushton when he said: "I do not have the space or the stomach to reply to all the points raised by Rushton" (Neisser 1997: 80). Yet, as a responsible psychologist who knows that this kind of dispute is ultimately resolved by empirical verification, he had no other choice but to concede that with respect to the racial differences in the mean measured sizes of skulls and brains, "there is indeed a small overall trend in the direction [Lynn and Rushton] describe" (ibid.). Apparently, this lesson in respect for hard empirical data has yet to be learned by many lovers of wisdom.

Philip Kitcher (2004: 13–14) recently replied to my criticism (which was first published in Sesardic 2000), but he only managed to dig himself into a deeper hole. Kitcher tried to justify his position by making the following two claims: (1) The fact that he ignored Michael's paper is not objectionable because virtually nobody saw that paper as a refutation of Gould, with the exception of the authors of *The Bell Curve* and J. Philippe Rushton, who – as Kitcher for some reason felt a need to stress – has "highly controversial views on race"; (2) Michael's measurements are not very strong evidence against Gould's views because Michael was a mere undergraduate student, whereas Gould was a professional paleontologist "whose own specialist work included some very meticulous measurements of fossil snails."

Let me briefly comment on both points.

(1) Why does Kitcher think that it is all right to ignore a paper if it is cited "only" by Herrnstein, Murray, and Rushton? Does he want to say that the works cited by scholars with "highly controversial views on race" may (or should) be completely disregarded? Well, in that case it would be useful if he could provide the full list of these "untrustworthy" authors so that we all know which bibliographic sources are suspect and not to be consulted. But more to the point, Kitcher is wrong that Michael's paper is cited against Gould "only" by the three authors he thinks he can dismiss so easily on ideological grounds. It is actually also mentioned in two recent books on human intelligence that are widely considered to be the best contemporary introductions to this research field. We can read there that Michael's paper "provides some corrections to Gould's account" (Mackintosh 1998: 234), and that "Gould's allegations of bias

were *refuted*" by Michael (Deary 2000: 265 – italics supplied, cf. 9). The good news is that these two texts are admissible on Kitcher's censorious criterion because neither Mackintosh nor Deary has "highly controversial views on race." It is somewhat odd that in his hunt for "untainted" references to Michael's article Kitcher searched the Internet but failed to look into two most obvious sources, and particularly Mackintosh's book, which I quoted in the very passage in which I criticized him.[12]

(2) Although Kitcher speaks about Gould's "measurements" and suggests that, because of Gould's rich paleontological experience and professional skills, his "measurements" are not less trustworthy than Michael's, the truth of the matter is that in this case Gould did no measurements at all. He only *reanalyzed* Morton's data and then concluded from this reanalysis that Morton's measurements must have been wrong and biased. Professor Janet Monge, Keeper of Physical Anthropology at the University of Pennsylvania Museum, says (personal communication): "We had never hosted Gould, and the Morton collection had been at the Museum at Penn since the early 1960s." It was Michael who did the measurements and he found no systematic error, no bias. Furthermore, it is unclear why Kitcher makes so much of the fact that Michael was an undergraduate student at the time when he published his article. The article passed the usual peer review process and came out in a leading anthropological journal. I would have thought that at that stage an author's rank and position in the academic pecking order become irrelevant, and that the quality of the argument is the only thing that counts. Kitcher should just think about what his own reaction would be if, by using the same reasoning, somebody downplayed the importance of his criticism of sociobiology (Kitcher 1985) on the grounds that he (Kitcher) was never even an undergraduate student of biology, whereas E. O. Wilson is a professional biologist. But Kitcher is wrong as well when he says that the disagreement between Gould and Michael is currently unresolvable without "further measurements" and "further analysis of the data." As I pointed out in my *Philosophy of Science* paper (Sesardic 2000), further measurements *had been made* a number of times (on samples other than Morton's collection), and the fact of a significant racial difference in brain size is today accepted

---

[12] Two other scholars from Kitcher's university also recently invoked Gould in rejecting Morton's conclusions and condemning them politically in no uncertain terms. But although they suggested that Morton was wrong in a way they regarded as "hideously immoral," they nevertheless referred the reader to Michael's paper "for a critique and reanalysis more favorable to Morton" (O'Flaherty & Shapiro 2004: 53). Ideology does not always trump scholarship.

in standard reference works[13] (see above). So, it is consistency with later measurements (conducted with better technologies and more precision) that makes it reasonable to believe that Michael was right and Gould wrong. Inexplicably, Kitcher did not address that part of my argument at all. Notice, however, that even if I were mistaken about this and if the jury were still out about who is right (Gould or Michael), my basic criticism of Kitcher would still stand. Namely, if experts in a given field cannot agree about what the basic facts are, a responsible scholar is not expected to report about these matters by mentioning only one side in the debate.[14]

1.5.3 The shadow of Cyril Burt

The case of British psychologist Cyril Burt is widely regarded as a dark episode in the history of hereditarianism. Burt's studies were once the cornerstone of the case for the high heritability of IQ, but nowadays his publications are no longer cited because of many serious doubts about his data collection and even his scientific integrity. Philosophers again proceed too quickly here. Eager to use the Burt scandal to further discredit the hereditarian approach, and apparently not interested in trying to understand what really happened in this complicated affair, they manage to get things wrong both about the historical context and about the scientific details of the story.

Kitcher starts his book about sociobiology (Kitcher 1985) with a grave warning about dangers of biological speculations about human psychology. His illustration of how hereditarianism can do social harm is the so-called "eleven-plus" exam, which for a number of years all children in Britain had to take at the age of eleven (including Kitcher himself, as he describes in a brief autobiographical aside). The purpose of the exam, which often involved an IQ test, was to select the most academically able children, who would then be accepted by schools with more demanding educational programs. Before telling a story about a girl who failed the exam, Kitcher claims that the educational division at age eleven

---

[13] A good source on these matters is Kitcher's own colleague at Columbia University, the physical anthropologist Ralph L. Holloway, who confirms (personal communication) that autopsy cases in general, and the collection at Columbia in particular, clearly show that Gould's denial of racial differences is empirically untenable.

[14] By the way, Gould also ignored Michael's contribution completely. Michael personally sent him his article but Gould didn't even mention it in the second edition of *The Mismeasure of Man*, despite its obvious relevance for the book's central theme.

was introduced under the influence of Cyril Burt's psychological theories about general intelligence, especially about its non-malleability. Luckily, he says, it is all now a thing of the past, but it shows how the whole educational system was wrecked by irresponsible psychological theorizing. Kitcher's diagnosis is endorsed by other philosophers (Dennett 1995: 483; Blackburn 2002: 30–31).

Although Dennett says that Kitcher's tale is "unanswerable," there is an easy answer: Kitcher is simply wrong about the facts. The introduction of the eleven-plus exam *had nothing to do with Burt or IQ testing.* The system actually started in the 1920s when secondary schools, which were until then fee-paying, opened up some free places. At first, students were chosen for these places in an ad hoc and chaotic way, sometimes on the basis of interviews with students (or even just with their parents) or exams in English and mathematics. It was only later that IQ tests were proposed and accepted as a more objective and fairer method of selection. So the educational division at age eleven, the main point of Kitcher's complaints, was already in place *before* IQ tests came on the scene:

> Britain's notorious "11+ examination" was *not a creation of psychometricians* led by Cyril Burt; it developed slowly out of the "free place examination" for grammar school scholarships, *instituted before any IQ tests existed.* The testers' growing influence produced the eventual inclusion of an IQ test and a new rationale for 11+, *but not the examination itself or its social functions.* (Samelson 1982: 656, italics supplied)

> Whatever Gould ... or Kamin ... may say, *Burt was certainly not responsible for the institution of the 11+ exam in English schools* after the 1944 Education Act, let alone for the practice of selection for "free places" in secondary schools in the 1920s and 1930s. Selection was already built into the system: the only question at issue was the basis on which it was to occur. (Mackintosh 1995a: 95, italics supplied)

> IQ tests were thus *not used to establish a system of selective secondary education.* That system was already in effect. (Mackintosh 1998: 24)

Notice that Mackintosh's correction of the widespread but factually mistaken claims about the 11+ exam starts with the words "Whatever Gould or Kamin may say." This suggests the most likely source that the philosophers relied on.

Also, despite all the outrage at the harm done by IQ tests, there is evidence that their effect was actually quite beneficial: "When the local education authority [in Hertfordshire] dropped IQ tests from its 11+ exams, there was an immediate and significant decrease in the proportion of children from working-class families entering grammar schools, and a comparable increase in the proportion of children from professional families" (Mackintosh 1998: 24).

So, contrary to what Kitcher says, IQ tests were *not* behind the introduction of the eleven-plus exam, and when IQ testing was eventually incorporated in the exam, it did *not* harm the underprivileged (under the circumstances as they were then, it actually opened up more educational opportunities for kids from poor families). It is odd that Kitcher, who describes how traumatized he was by having to take the eleven-plus exam himself, showed no curiosity to look into the matter more deeply and learn about the *real* historical roots and rationale for that educational reform.

Simon Blackburn criticizes Steven Pinker because in his discussion of studies of identical twins (Pinker 2002), "he does not refer to Cyril Burt, the British psychologist who wrecked the education system on the basis of such evidence [studies of identical twins], having made it all up" (Blackburn 2002: 29–30). This half-sentence has two problems.

First, as we saw, it is not true that Burt wrecked the education system. Second, the claim that Burt "made up" the evidence in his twin studies is just one opinion in the ongoing debate about the whole affair.[15] The fraud theory was, for example, strongly disputed in two books devoted to the Burt case (Joynson 1989; Fletcher 1991) and in several scholarly articles. Even Mackintosh, who enjoys the reputation of a fair and impartial commentator, concluded his detailed analysis of the topic in this way: "If the MZ twin data were the only grounds for accusing Burt of fraud, one would probably have to give him the benefit of the doubt" (Mackintosh 1995b: 68). Blackburn's resolute pronouncement on an issue that is still under debate (see also Ward 1998; Joynson 2003) shows the *modus operandi* of the anti-hereditarian consensus: an argument of one side is repeated so often that it comes to be regarded as the truth, and there is no evidence of awareness at all of the existence of alternative opinions.[16]

---

[15] Blackburn does not seem to be aware that with his assertion that Burt's data were all made up, he has a problem of how to explain the fact that some of the best contemporary studies agree quite well with these alleged total fabrications. For example, recently several behavior geneticists laughingly said that they were "grateful as hell" that in their study of monozygotic twins reared apart, the IQ correlation turned out to be .78, and so mercifully not bang identical to the infamous magnitude of .77, reported by Burt (Miele 2002: 99–100).

[16] Contrast Blackburn's precipitate judgment (shared by Goldman 1999: 34) with Susan Haack's judicious comment: "That Burt was a fraud came to be taken as established fact

1.5.4 *Racism* ex nihilo

One of the rare cases in which a philosopher of biology devoted an entire text to discussing heritability is Sahotra Sarkar's chapter "Obsessions with heritability" in his book *Genetics and Reductionism*. The result is disappointing and further confirms the grim picture. After rehearsing the old arguments against the use of heritability (based on the possibility of statistical correlation and statistical interaction), Sarkar clinches his attack with a standard speculation about motives: "It is hard, therefore, not to suspect that the continued pursuit of [heritability] is guided, at least to some extent, by non-cognitive, especially political, factors" (Sarkar 1998: 92). He mentions three sets of considerations that point to that conclusion, all of which are utterly inadequate to establish such a sweeping political accusation. I will comment only on two of them.

First, Sarkar infers political motivation from the fact that "the traits for which [heritability] continues to be pursued often include those carrying social judgments, even if they are ill-defined" (Sarkar 1998: 93). After giving examples of religiosity and IQ, Sarkar continues:

> Bouchard . . . reports relatively high values of [heritability] for "openness," "agreeableness," "conscientiousness," "neuroticism," and "extroversion," *each of which is a trait that carries social judgment.* (Sarkar 1988: 93 – italics added)

Sarkar's point is very clear: the continued pursuit of heritability must be politically motivated because the traits picked out for heritability research often include those carrying social judgment, like the five traits mentioned. Indeed, why were exactly these five traits singled out for heritability studies?

One possibility is, as Sarkar suggests, that some people had a sinister political intention to take traits "that carry social judgment" and then perform a heritability analysis in order to develop the hereditarian argument and justify the oppression of certain social groups. But acquaintance with some very basic psychology points to a much simpler and quite benign explanation for the choice of the quintuple. Namely, these five traits are known in psychology as *the five main personality traits*. If one happens to know this, it becomes quite obvious that these traits were put in the

after his biographer, Leslie Hearnshaw . . . endorsed it. [Later], however, Robert Joynson argued that Burt was guilty of nothing worse than the occasional carelessness. *Now, after reading Joynson's book, I can honestly say I don't know*" (Haack 2003: 200 – italics added).

foreground not by Bouchard, but by the wide consensus of personality psychologists. And then, there is absolutely no need to invent a right-wing conspiracy of scientists in behavior genetics. The traits in question became salient in psychology simply because they emerged as very robust results of the systematic empirical research on human personality. (Needless to say, one could criticize the way these traits were picked out as "the big five," but that would be a completely different topic, unrelated to discussions about heritability.)

Second, Sarkar fortifies his imputation of political motives by introducing another consideration as well. He says that it is "hard not to suspect" that research on heritability is guided by political factors because "the work on [heritability] and IQ has been routinely used to argue for genetic inferiority of certain groups, particularly African-Americans" (1998: 92–93). There is no argument here at all. Sarkar hypothesizes (without offering any evidence) that scholars who accept the genetic explanation of the racial IQ difference entered this research area just because they wanted to give an aura of scientific respectability to their racist prejudices. But ironically, Sarkar himself provides the best refutation of his own claim. Speaking about those authors who advocated the genetic account of the IQ differences between ethnic groups, Sarkar gives reference to Richard Herrnstein's article "IQ" from *Atlantic Monthly* (Herrnstein 1971) and his book *IQ and Meritocracy* (Herrnstein 1973). In reality, however, Herrnstein at that time did *not* subscribe to the genetic hypothesis about white–black IQ variation. In the early 1970s he disagreed with Jensen, and it was *only later* that he changed his mind, and was converted to hereditarianism with respect to racial differences. In the book *IQ and Meritocracy*, Herrnstein complained about frequent misinterpretations of his views, and distanced himself from Jensen very explicitly:

> My article took what might be called an explicitly agnostic stand on racial (i.e., black white) differences in tested intelligence . . . I believe that racial and ethnic group differences are hard to pin down as regards inheritance. My interest was not race, but social class differences. (Herrnstein 1973: 12)

As one of the authors of *The Bell Curve* (Herrnstein & Murray 1994), Herrnstein has become notorious for his views on race, genetics, and IQ. This may be a reason why many people who did not take the trouble to study the sources tend to believe that he must then have defended the very same ideas in his writings from the early 1970s, which gave rise to a heated political controversy too. But although this mistake is not uncommon (see Etzioni 1973: 112; Nelkin & Lindee 1995: 114; Segerstrale 2000: 270, 283) and is in a sense expected from lay readers, it is disturbing when it comes from a philosopher of science writing about heritability in a high-profile philosophical publication.[17]

Another philosopher who made the same mistake is Michael Dummett (Dummett 1981: 295–296). He also accused Herrnstein of racism, although at that time Herrnstein had resolutely refused to take a stand on issues involving race.

---

[17] Surprisingly, despite the non-existence of support for these ideological accusations, Samir Okasha says that Sarkar's conclusion about political motivation driving heritability research is "based on a meticulous and technically expert critique of [the] underlying methodology" (Okasha 2000: 183).

2

A tangle of interactions: separating genetic and environmental influences

> No aspect of human behavior genetics has caused more con-
> fusion and generated more obscurantism than the analysis
> and interpretation of the various types of non-additivity and
> non-independence of gene and environmental action and
> interaction.
>
> Lindon J. Eaves

2.1 TWO CONCEPTS OF INTERACTION

A widespread conviction that heritability claims are devoid of almost any interesting explanatory content is often based on an argument that genes and environments *interact*, and that for this reason their causal contributions to phenotype cannot be separated and measured independently. An immediate problem with this argument is that there are two very different meanings of "interaction": commonsense and statistical. According to the commonsense notion (interaction$_c$), to say that two causes A and B interact means that neither can produce the effect without the presence of the other. To use a standard example, striking a match and the presence of oxygen interact to produce fire. According to the statistical notion (interaction$_s$), however, to say that two variables A and B interact means that a change in one variable does not always have an effect of the same magnitude: its effect varies, depending on the value of the other variable. For instance, the very same life event, such as parental divorce, may affect children with different personality characteristics quite differently.

Already a half a century ago Waddington proposed that, in order to avoid confusion, the term "gene–environment interaction" should be reserved for the statistical concept:

> This expression [gene–environment interaction] is derived from statistical terminology and, as this example makes clear, is used in a much more restricted sense than might appear at first sight. It is to be avoided when one wishes merely to indicate that the phenotypic effects of a genotype are influenced by the environment, in order that it can be restricted to its special use to designate cases in which phenotypic effects of different genotypes are differently affected by a given environmental change. (Waddington 1957: 94)

As usually happens with such attempts at terminological legislation, Waddington's intervention did not have much effect. The equivocation continued. This is unfortunate because it does tend to generate confusion in the following way. Interaction$_c$ of genes and environments is always present but it generates no problems for the estimation of heritability. On the other hand, the existence of strong interaction$_s$ between genes and environments may really undermine the usefulness of heritability claims, yet the existence of such interaction is itself an open empirical question. Briefly, interaction$_c$ is ubiquitous but irrelevant for discussions about heritability, whereas strong interaction$_s$ is potentially a problem for heritability, but the extent of its presence remains a contentious issue. Nevertheless, due to the equivocation, a quick and fallacious argument against heritability is occasionally developed. The argument (reconstructed from Vreeke 2000a: 54; Wahlsten 2000a: 46; Meaney 2001: 52) goes like this:

(1) Heritability claims are based on the assumption of additivity.
(2) Additivity means that genes and environments act separately (i.e., do not interact).
(3) Obviously, genes and environments do not act separately (i.e., they interact).

Therefore:

(4) Heritability claims are based on a false assumption.

The argument trades on the ambiguity between interaction$_c$ and interaction$_s$. Statement (2) is true in the sense that additivity is indeed incompatible with some forms of strong interaction$_s$. Statement (3) is true in the sense that interaction$_c$ between genes and environments is empirically indisputable. But if we consistently stick to one meaning of

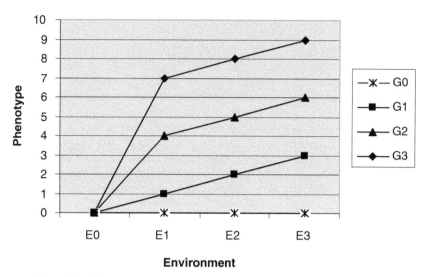

Figure 2.1 Commonsense interaction – yes, statistical interaction – no.

"interaction," the two statements cannot both come out true. If statement (2) is interpreted as saying that additivity is incompatible with interaction$_c$, it is false. Similarly, (3) is false if it says that interaction$_s$ of genes and environments is a matter of obvious truth.[1]

Figure 2.1 is supposed to clarify things by showing a situation where there is interaction$_c$ but no interaction$_s$. If zero indicates non-existence, then the trivial fact of interaction$_c$ (namely, that without either genotype or environment there is no organism) is illustrated by the phenotype being zero for the zero genotype (whatever the value of the environment), and also for the zero environment (whatever the value of the genotype). However, despite interaction$_c$, there is no interaction$_s$ because as long as neither variable has zero value, the contribution of environment is always the same (whatever the genotype), and the contribution of genotype is always the same (whatever the environment).

The absence of statistical interaction is reflected in the parallel lines of the three genotypes (G_1, G_2, and G_3) in the environmental range from E_1 to E_3. This shows perspicuously that moving from one environment to another always produces the same phenotypic change (independently of

---

[1] By the way, I disagree with (1) as well. Heritability claims do *not* presuppose additivity. Or, to put it differently, it makes sense to talk about heritability even when there is no additivity at all. In cases of complete non-additivity (i.e., where the entire variance is due to statistical interaction between genes and environments), heritability is simply 0.

the genotype), and that moving from one genotype to another always produces the same phenotypic change (independently of the environment).

Strictly speaking, there is a whiff of statistical interaction in Figure 2.1 because the genotype lines are not parallel throughout the whole range. The extreme cases of no genotype (G_0) and no environment (E_0) destroy the parallelism, because in these cases any change in the other variable has zero effect (instead of the effect of its usual magnitude). However, if we ignore these extreme cases and just concentrate on the situations where organisms exist (i.e., where phenotypes have non-zero values), the parallelism reigns supreme and statistical interaction completely disappears.[2]

2.2 THE RECTANGLE ANALOGY

Let us consider a situation where statistical interaction makes it impossible to assign separate causal contributions to genes and environments.

For simplicity, take a population of organisms (see Figure 2.2) with two different genotypes, G_1 and G_2, that are with equal frequency distributed in two different environments, E_1 and E_2. In environment E_1 organisms with genotype G_2 have higher phenotypic value (with respect to some trait P) than organisms with genotype G_1; in environment E_2, on the other hand, it is G_1 organisms that have higher P value. Now, should the phenotypic differences in that population be ascribed to genes or to environment? Either answer is plainly wrong. Assuming that genotypes and environments are uniformly distributed, neither the genetic difference (between G_1 and G_2) nor the environmental difference (between E_1 and E_2) has a net effect on P: their average effect is clearly zero. The reason for this is that the effect of genotype is opposite in the two environments, and the effect of environment is also opposite for the two genotypes. Therefore, phenotypic variance is here due neither to genes nor to environment alone: it is the outcome of statistical interaction between genes and environments.

The change in factor G from G_1 to G_2 increases phenotypic value P if factor E has value E_1, whereas it has negative effect otherwise (if E is fixed on E_2). Under these circumstances it hardly makes sense to ask what is the effect of G *tout court*. It depends.

---

[2] Recall, the point of this example is not to show that parallelism is frequent or typical, but just that interaction$_c$ does not necessarily imply interaction$_s$.

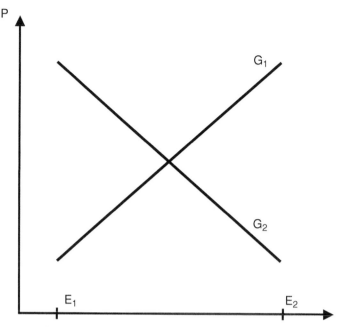

Figure 2.2 Statistical interaction: No Main Effects.

In the case pictured in Figure 1.1, however, we could easily divide phenotypic variation into genetic and environmental parts. In that case, moving from one environment to the other brought about the same phenotypic change, irrespective of genotype. Similarly, shifting from one genotype to another had always the same phenotypic effect (in both environments). Environmental and genetic contributions to variance added up to total variance: $V_P = V_G + V_E$. This neat solution is in marked contrast to the intractable case from Figure 2.2 where norms of reaction cross and where the effects of genes and environment become hopelessly jumbled.

It would be a serious mistake, though, to conclude on the basis of Figure 2.2 that the presence of statistical interaction *always* precludes the existence of main effects. The situation depicted in Figure 2.2 is an extreme case of statistical interaction, and no general conclusions should be hastily drawn from it. There are moderate forms of statistical interaction that leave a lot of space for the operation of main effects.

That there is not enough awareness of this simple point is attested in the frequently made claim that the nature–nurture discussion is as meaningless as asking whether the area of a rectangle is more determined by its length or its width (Cosmides & Tooby 1997; Hebb 1980: 72; Johnston

2003: 99; Ehrlich & Feldman 2003: 102; Lerner 1986: 85; Ehrlich 2000; Herbert 1997; Meaney 2001: 51; Eisenberg 1995: 1571; 2001: 378; 2002: 339). This notorious rectangle analogy is problematic for at least four reasons. (1) It presupposes the existence of statistical interaction between genes and environments, thus prejudging a controversial issue. (2) It assumes (wrongly) that the non-linear dependence of the area of a rectangle on its length and width makes it impossible to separate and measure the contributions of these two factors. (3) By focusing on an individual rectangle, the analogy misrepresents the nature–nurture debate, which is best construed as being about population differences, not about an individual case. (4) Finally, even when talking about an individual case, the question whether a particular person's phenotypic deviation from the mean is more due to genes or environment is not necessarily meaningless. Let me briefly explain each of these four points.

(1) The relation of the area of a rectangle to its length and width is multiplicative: area = length × width. Those who offer the rectangle analogy usually supply no evidence (or no convincing evidence) for the belief that heredity–environment relation is generally non-linear. This is a completely arbitrary assumption on their part. By the way, it should be stressed that in some situations a non-linear relationship can be turned into an additive one by a simple scale transformation. In the rectangle case, the multiplicative nature of the relationship (area = length × width) is readily changed into additivity by a logarithmic transformation: log(area) = log(length) + log(width).

(2) Look at Figure 2.3 to see that even in cases of non-linear causal dependence it may still be possible to separate and quantify the contributions of the statistically interacting causes.

Here we have an example with nine different rectangles, which include all possible combinations of three different widths (1, 2, and 3) and three different lengths (1, 50, 100). Statistical interaction is conspicuous: the three lines, representing the changing areas of rectangles with different lengths, are *not* parallel. Nevertheless, it is easily seen from the graph that, on average, a vertical change (a change in length) produces a stronger effect on the area than a horizontal change (a change in width). In other words, the variation in rectangle areas is more influenced by length differences than by width differences. Moreover, if we want numerical precision, a simple ANOVA (analysis of variance) calculation tells us that the total area variance (9312) can be decomposed into three parts: width variance (1687), length variance (6495), and variance due to width–length interaction (1130). So, the contribution of differences of lengths to the

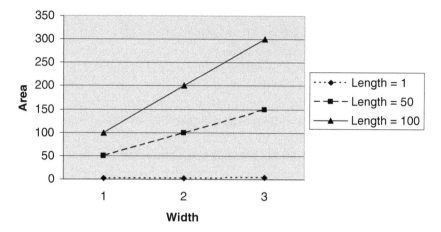

Figure 2.3 Rectangle area, length, and width.

area variation is around 70 percent while the two remaining sources are considerably less important. The width contribution is 18 percent, and the interaction of length and width accounts only for 12 percent of the entire area variance.

The point of this simple and artificially constructed example is not to argue that statistical interaction *always* leaves room for the existence of meaningful main effects. Rather, the purpose is to show that the existence of statistical interaction does not exclude *a priori* the possibility of main effects.[3]

(3) Raising a question about how the area of an *individual* rectangle is influenced by its width and length amounts to abandoning a *population* perspective. With this move the connection with the nature–nurture controversy becomes tenuous because that debate makes most sense as looking for answers at the level of population differences.

(4) It is often said that in individual cases it is meaningless to compare the importance of interacting causes: "If an event is the result of the joint operation of a number of causative chains and if these causes 'interact' in any generally accepted meaning of the word, it becomes conceptually impossible to assign quantitative values to the causes of that *individual event*" (Lewontin 1976a: 181). But this is in fact not true. Take, for example, the rectangle with width 2 and length 1 (from Figure 2.3). Its area is 2,

---

[3] So it is not true that statistical interaction is "the bane of heritability analysis" (Wahlsten & Gottlieb 1997: 176), or that "nonadditivity of the contributing causes makes it invalid to partition the variance into distinct components and thereby renders heritability coefficients meaningless" (Wahlsten 1990a: 110).

which is considerably below the average area for all rectangles (around 100). Why is that *particular* rectangle smaller than most others? Is its width or its length more responsible for that? Actually, this question is not absurd at all. It has a straightforward and perfectly meaningful answer. The rectangles with that width (2) have on average the area that is identical to the mean area for all rectangles (100.66), so the explanation why the area of that particular rectangle deviates so much from the mean value cannot be in its width. It is its below-average length that is responsible.

Even the usually cautious David Lykken slips here by condemning the measurement of causal influences in the individual case as inherently absurd: "It is meaningless to ask whether Isaac Newton's genius was due more to his genes or his environment, as meaningless as asking whether the area of a rectangle is due more to its length or its width" (Lykken 1998a: 24). Contrary to what he says, however, it makes perfect sense to inquire whether Newton's extraordinary contributions were more due to his above-average inherited intellectual ability or to his being exposed to an above-average stimulating intellectual environment (or to some particular combination of the two). The Nuffield Council on Bioethics makes a similar mistake in its report on genetics and human behavior: "It is vital to understand that neither concept of heritability [broad or narrow] allows us to conclude anything about the role of heredity in the development of a characteristic in an individual" (Nuffield 2002: 40). On the contrary, if the broad heritability of a trait is high, this does tell us that any individual's phenotypic divergence from the mean is probably more caused by a non-standard genetic influence than by a non-typical environment. For a characteristically clear explanation of why gauging the contributions of heredity and environment is *not* meaningless even in an individual case, see Sober 1994: 190–192.

Cosmides and Tooby also argue that a heritability value tells us nothing about an individual:

A heritability coefficient measures sources of *variance* in a *population* (for example, in a forest of oaks, to what extent are differences in height correlated with differences in sunlight, all else equal?). It tells you nothing about what caused the development of an *individual*. Let's say that for height, 80% of the variance in a forest of oaks is caused by variation in their genes. This does not mean that the height of the oak tree in your yard is "80% genetic." (What could this possibly mean? Did genes contribute more to your oak's height than sunlight? What percent of its height was caused by nitrogen in the soil? By rainfall? By the partial pressure of CO_2?) When applied to an individual, such percents are meaningless, because all of these factors are necessary for a tree to grow. Remove any one, and the height will be zero. (Cosmides & Tooby 1997)

They criticize some manifestly absurd ways of applying a population percentage to an individual, but they never consider a quite simple population-to-individual inference that is not "meaningless" at all. In fact, Cosmides and Tooby are wrong that heritability tells us *nothing* about what caused the development of an individual. If the heritability of a normally distributed trait in a given population is .8, this does tell us that the deviation of an arbitrarily chosen individual organism from the phenotypic mean is in all likelihood caused more by its deviation from the genetic mean than by its deviation from the environmental mean.

Someone may object (as did a reader of an earlier draft of this chapter) that an individual phenotypic deviation from the group mean should not "really" count as a fact about an individual. Being a *relational* fact, and *not* an *intrinsic* fact about an individual, it is not information "about an individual" in the natural sense of that phrase. I disagree. I think that many population-relative facts are usually regarded as giving important information about individuals, like being an Olympic champion, having an IQ above 140, being a best-selling author, being a surgeon with an unusually high patient mortality, etc.

2.3 LEWONTIN AGAINST ANOVA

The best known anti-hereditarian argument based on statistical interaction of genes and environment is presented in Richard Lewontin's paper "The Analysis of Variance and the Analysis of Causes" (Lewontin 1976a). It was first published in the *American Journal of Human Genetics* in 1974, but I quote it from the version reprinted in Block and Dworkin's anthology because this book is more accessible. The article made a big impact on later discussions of these issues (see Block & Dworkin 1976b: 533; Layzer 1976: 201; Sober 1984: 107–108 and 266; Kitcher 1985; Gifford 1990: 328; Wahlsten 1990a: 110; Oyama 1992; Griffiths & Gray 1994: 304; Wahlsten 1995: 253; Berkowitz 1996; Godfrey-Smith 1998: 51; Kitcher 1999: 89; Sterelny & Griffiths 1999: 16–17; Godfrey-Smith 2000: 27; Kaplan 2000: 38–42; Oyama 2000a: 39, 107; Vreeke 2000b: 38–39; Falk 2001: 132–134; Gray 2001: 187; Kitcher 2001b: 396, 412; Oyama et al. 2001: 3; Pigliucci 2001: 58–65; Godfrey-Smith 2002: 587; Johnston & Edwards 2002: 31;

Robert 2003: 976; Wahlsten 2003: 21; Downes 2004; Maclaurin 2002: 116–121; Garfinkel 1981: 119).

Although Lewontin's article is called a "landmark paper" (Pigliucci 2001: 58), "classic discussion" (Godfrey-Smith 2000: 27), "classic essay" (Kitcher 1989: 264), "*locus classicus*" (Okasha 2003), "seminal paper" (Griffiths & Knight 1998: 257; Pigliucci 2001: 65), "key article" (Schaffner 2001: 488), "brilliant analysis" (Gray 2001: 187), or even "the single most influential contribution to the literature on the interpretation of behavioral genetics" (Griffiths 2002a), it is not easy to see what its main contribution really is. Its importance is sometimes ridiculously overstated, as when Pigliucci, for example, says that in that paper Lewontin was "the first to suggest that heritabilities can change depending on the environment" (Pigliucci 2001: 65). Needless to say, the environment dependence of heritability logically follows from its being a population statistic, and it is ludicrous to suggest that quantitative geneticists were totally unaware of this trivial fact[4] until Lewontin "discovered" it in 1974.

Before I go to the main argument of Lewontin's paper, let me briefly discuss a powerful analogy that he uses to illustrate the absurdity of separating the influences of heredity and environment:

> if two men lay bricks to build a wall, we may quite fairly measure their contributions by counting the number laid by each; but if one mixes the mortar and the other lays the bricks, it would be absurd to measure their relative quantitative contributions by measuring the volumes of bricks and of mortar. It is obviously even more absurd to say what proportion of a plant's height is owed to the fertilizer it received and what proportion to the water, or to ascribe so many inches of a man's height to his genes and so many to his environment. (Lewontin 1976a: 181–182)

There is no doubt that of the two situations, "brick–brick" and "brick–mortar," the nature–nurture problem much more resembles the latter. In the brick–brick situation, either of the two workers could build the wall alone. But obviously genes and environments cannot build an organism

---

[4] But, as we have learned by now, if something is a matter of elementary knowledge in biology it is not necessarily so in philosophy of biology. John Dupré, for example, says: "The heritability of a trait . . . is *affected only by phenomena at one structural level, the genetic* . . . different samples of a population can be expected to reveal approximately the same trait heritabilities, since *these will not depend on environmental variables*" (1995: 137 – italics added). Dupré's mistake is hard to explain. Since heritabilty is the *proportion* of the entire variance that is due to genes, it is evident that as the environmental contribution to the total variance increases, the proportion of genetic influence (heritability) will decrease, and vice versa (even if nothing changes at the genetic level). Heritability *does* crucially depend on environmental variables.

in isolation from one another. *Both* are needed. So, since the heredity–environment problem is more similar to the brick–mortar situation, and since in that situation it is absurd to quantify the contributions of bricks and mortar, Lewontin wants us to conclude that it is also absurd to quantify the contributions of heredity and environment. But this is too quick. The fact that the heredity–environment indeed resembles the brick–mortar situation more than the brick–brick situation does *not* mean that it resembles the brick–mortar situation *enough* to allow inferences from one case to the other.

There are two important differences. First, the heredity–environment problem is about *populations*, whereas the brick–mortar situation is about an *individual* wall. And second, in the brick–mortar situation, what is claimed to be absurd is measuring the relative quantitative contributions of the two men by measuring the *volumes* of bricks and of mortar. There is no analogy to the heredity–environment problem in this respect because no one wants to measure the relative quantitative contributions of genes and environments by measuring the *quantities* of genes and environments. Rather, the idea here is to measure the effects of different *kinds* of genes and environments.

Now after locating the two crucial differences that show the weakness of Lewontin's analogy, let us modify the brick–mortar situation in these two respects in order to make it as similar as possible to the heredity–environment problem. This is a very illuminating exercise because we will see that in the reconstruction of the brick–mortar situation that keeps relevant similarities with the nature–nurture problem, quantifying the relative contributions of bricks and mortar is no longer absurd.

First, instead of discussing an *individual* wall, we will discuss a *population* of walls. And second, instead of trying to measure the contributions of different *volumes* of bricks and mortar, we will attempt to measure the contributions of different *kinds* of bricks and mortar. Suppose there are three kinds of bricks (B_1, B_2, and B_3) and three kinds of mortar (M_1, M_2, and M_3). We want to measure the contributions of these different kinds of bricks and mortar to the stability of walls that arise out of nine different B–M combinations. Here is the translation manual for the analogy: bricks are genes, mortar is environment, and wall stability is phenotype.

Figure 2.4 shows a hypothetical distribution of stability values for nine walls (all possible B–M combinations). Now, does it make sense here to quantify the contributions of bricks and mortar? Yes, it does. We see from the graph that switching from one kind of brick to another makes a much bigger difference to the wall stability than changing the kind of

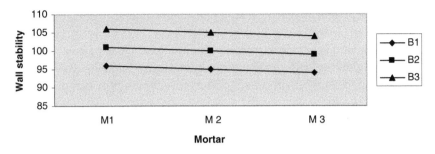

Figure 2.4 Bricks and mortar.

mortar. The main effect of B is 5 and the main effect of M is 1. If we want further numerical precision, a simple ANOVA calculation tells us that the variation in stability of these walls is 97 percent due to the brick variation and only 3 percent to the mortar variation. The relation between the two causes is completely additive, and there is no brick–mortar interaction (in the statistical sense).

I deliberately chose an example with an extremely strong influence of one factor in order to show that, despite both causes being causally necessary for the effect, their contributions to the population variance can be measurable, and there is no guarantee that these contributions will be even remotely equal. The general moral: when the brick–mortar situation is presented in the way that makes it relevantly similar to the heredity–environment problem, there is nothing absurd in the attempt to quantify causal contributions of the two factors. Notice also that once the population perspective is in place, it becomes possible to quantify the contributions of the two factors even in *individual* cases. For instance: the wall made of bricks B_1 and mortar M_1 has a below-average stability exclusively because of the below-average quality of its bricks.

Enough of bricks and mortar, and back to the central argument from Lewontin's paper. What he tries to show was that, contrary to widespread opinion, the analysis of variance cannot bring us knowledge about the extent to which genes causally determine phenotypic differences. Why not? Well, here is a three-step structure of his argument, extracted from the original text by careful reading and a sincere attempt to make the strongest case for his views. (I invite the readers to check the source and satisfy themselves that my interpretation is fair and that it isolates the main thread of the argument.)

(1) *Non-additivity*. Genes and environment typically interact (in the statistical sense) to produce phenotypes.

(2) *Locality*. Because of (1), a high heritability value obtained in a given range of environments may be just the local characteristic of environmental and genetic influences within that particular narrow range of environments, with the situation dramatically changing over a more extended set of environments. Therefore, heritability claims are not generalizable.

(3) *Causal irrelevance*. Because of (2), analysis of variance is useless as an instrument for gaining etiological knowledge, in that we have no reason to believe that it will give us a correct picture of the *general* relationship between cause and effect.

Let us take a closer look at each of these three steps.

2.4 NON-ADDITIVITY

Lewontin's criticism of ANOVA is often presented as making a purely theoretical point about inherent limitations in any attempt to derive causal conclusions from statistical data by using the analysis of variance. This interpretation is supported by the way he himself describes the upshot of his argument at the outset: "I will begin by saying some very obvious and elementary things about causes, but I will come thereby to some very annoying conclusions" (Lewontin 1976a: 180). However, a brief look into the argument is enough to show that the quoted description is incorrect. The "annoying" conclusions, far from being derived merely from "some obvious and elementary things about causes," are in reality obtained only with the help of an additional, *contingent* (and, as we will see, highly controversial) empirical premise, namely, that strong statistical interaction between genes and environments is rampant. Lewontin concedes twice (Lewontin 1976a: 184, 189) that there are possible empirical situations in which his objections to causal implications of heritability do not work. These are the situations where statistical interaction between genes and environments is either non-existent or present just to a mild degree.

So how does Lewontin know that non-additivity prevails over additivity? There are two things in his article that can be seen as an attempt to establish the domination of non-additivity: first, the choice of graphs representing various possibilities, and second, empirical evidence.

There are eight graphs in Lewontin's article, illustrating some of the possible relations between genotypes, environments, and phenotypes. Interestingly, most of these graphs (six out of eight) represent especially strong cases of interaction, so-called "non-ordinal" interactions, in which the norms of reactions of two genotypes cross each other and in which it becomes difficult to talk about main effects (genetic and environmental). In the remaining two cases, in which the norms of reaction of two geno-types are parallel or nearly parallel, the considerations that Lewontin uses for mounting a methodological attack on heritability inferences are "irrelevant," by his own admission (Lewontin 1976a: 189). Now, although the crossing norms of reactions so strongly dominate Lewontin's selec-tion of examples and although he calls the situation of additivity or near-additivity "a very special case" (Lewontin 1976a: 184), it is clear that the issue cannot be resolved *a priori*, by a tendentious choice of examples.

James Maclaurin falls into a trap here. Relying on Lewontin, he states that "in *most* cases, genes and environment are not independent causes of phenotype" (Maclaurin 2002: 119 – italics added), and that "perfect additivity is the exception, not the rule" (Maclaurin 2002: 119), but the only consideration he gives in support of these claims is that just one of Lewontin's eight graphs (all duly reproduced in Maclaurin's paper) illustrates perfect additivity, whereas "all the other graphs represent a lack of additivity." Well, perfect additivity is indeed an exception *among Lewontin's diagrams*, but this hardly constitutes an adequate reason to believe that it is also an exception *in reality*. Whether it is or not is an empirical question, which Maclaurin answers in a non-empirical way (i.e., by looking at Lewontin's pictures, instead of looking at the world).

All this reveals that the impact of Lewontin's criticism of behavior genetics crucially depends on how much the additivity of genetic and environmental influences on human phenotypes is *empirically* violated. For to the extent that additivity is even approximately preserved, his argument loses most of its force. Given the absolutely essential role that the empirical premise plays in his argument, it is surprising how little evidence he actually provided in its support.

In two brief paragraphs specifically devoted to empirical evidence Lewontin (1976a: 190–191), says three things: (a) measurement of human norms of reaction for complex traits is impossible "because the same geno-type cannot be tested in a variety of environments"; (b) even in research on animals and plants (where experiments are possible) very little work has been done to characterize the norms of reaction; (c) in one study on plants and one on Drosophila strong G–E interaction was discovered. Each of these claims raises serious questions that, to my knowledge, have never been asked before.

The strange thing about (a) is that, despite defending an extreme methodological claim that measurement of human norms of reaction is *impossible*, Lewontin neither discusses nor so much as mentions what were then most important contributions to the literature on that issue. At the time, the most sophisticated analysis of methodological problems in human behavior genetics was undoubtedly the paper by Jinks and Fulker (1970), which was an important step in the development of powerful model-fitting methods that dominate the contemporary scene. Moreover, John Jinks and David Fulker did suggest how G–E interactions could be empirically detected *without* testing the same genotype in a variety of environments (see below), which directly contradicted Lewontin's impossibility claim. It is hard to explain why Lewontin fails to address their argument, or at least inform the reader about this "landmark paper" (Neale & Cardon 1992: 31).[5] If somebody thinks that Jinks and Fulker's article was too recent and maybe for that reason unknown to Lewontin (it was actually published *four* years before Lewontin's paper), there are still some older and highly relevant works that Lewontin also completely ignores. What first comes to mind is Raymond Cattell's attempt to approach G–E interactions by using his method of multiple abstract variance analysis (MAVA), also a crucial precursor of the current research techniques in behavior genetics (Cattell 1963; Cattell 1965).

As for (b), the fact that there has been *very little* work on animal norms of reaction is actually a reason *against* accepting Lewontin's belief about the omnipresence of interaction. It rather supports agnosticism.

As for (c), it is interesting that, by way of empirical evidence, Lewontin cites only two studies, and that even these two studies are not presented as being representative or giving typical results, which is the only way they could support his strong conclusion. Instead, the study of plants is introduced as "the classic work," and the study on Drosophila is described as "*an* example of what has been done in animals" (Lewontin 1976a: 191 – italics added).

Even in a recent article written for the *Stanford Encyclopedia of Philosophy* (Lewontin 2004), with all the empirical evidence accumulated in the thirty years since 1974, Lewontin's premises are still insufficient to

---

[5] Even Michael Lerner, who usually sees eye to eye with Lewontin over many issues about human behavior genetics, says: "In any case, the paper of Jinks and Fulker is a must for any serious student of the subject" (Lerner 1972: 408).

establish the conclusion that he wants his readers to draw. When he says that "the mapping of different genotypes into phenotypes in one environment is *often* completely unpredictable from their mapping in another environment" (italics added), the word "often" is clearly too weak to justify the claim of *ubiquitous* non-additivity. When he says that "*many* experiments on *many* different organisms . . . show this same result" (italics added), the word "many" allows for the possibility that *much more* experiments on *much more* different organisms do *not* show the same result. When he says that "it is *the common* experience that norms of reaction of different genotypes are curves of irregular shape that cross each other" (italics added), it leaves open the possibility that non-crossing norms of reaction are in fact a *much more common* experience.

Had Lewontin formulated his premises more strongly, they would have supported his conclusion. But then the big question would be how to show that they are really *true*. The advantage of the current phrasing is that Lewontin is on pretty safe ground with his premises, and he also *seems* to advance his conclusion. Yet more careful readers will notice that such weak premises cannot justify the intended inferential move.

Despite all this being a manifestly poor ground to support the sweeping claim of pervasiveness of non-additivity in biology, Lewontin's paper has been nevertheless widely regarded as establishing precisely that. For example, Russell Gray thinks that it is "very clear" that the relationship between genotype and environment is "strikingly non-additive" (Gray 1992: 174), but his sweeping generalization about biological development is backed up only by *one* empirical study of plants, the same one that has been repeatedly cited by Lewontin (1976a: 191; 2000: 21; 2004; Feldman & Lewontin 1976: 14), and which, by the way, was conducted about seventy years ago.

Block and Dworkin show a graph with crossing norms of reaction of two genotypes (see Figure 2.5), and maintain that this type of case is "not atypical" (Block & Dworkin 1976b: 483–484). They give no sources to support this claim but in a note to the same paragraph they say that they "are indebted to [Lewontin 1976a] for many of the points in the rest of this section" (Block & Dworkin 1976b: 533). Did they have good grounds to think that their example was *not* atypical? Well, let us see. At about the same time that Block and Dworkin published their piece, a group of leading quantitative geneticists declared that "a trait was atypical if more than about 20 per cent of the measured variation could be attributed to G × E [interaction]" (Eaves et al. 1977: 3). I calculated the variances

63

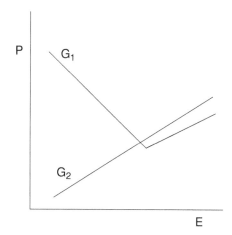

Figure 2.5 A "not atypical case"? (From Block & Dworkin 1976b: 484.)

for the Block–Dworkin graph (Figure 2.5),[6] and it turned out that the contribution of interaction in that situation was around 75 percent!

When Sterelny and Griffiths insist that gene–environment interactions "are typically not additive" (Sterelny & Griffiths 1999: 16), they exclusively rely on Lewontin's article. Griffiths also says that in that paper "Lewontin pointed out that the *empirical evidence* suggests that actual norms of reaction are likely to be non-additive" (Griffiths 2002a – italics added). Two things are strange here.

First, as we saw, Lewontin did not give much empirical evidence to support his views. Moreover, in a later article he himself admitted that except for one of the studies he cited in 1976a, "no study of the norms of reaction of naturally occurring heterozygous genotypes [had] appeared until [1982]," and that the basic data for judging how particular genotypes react to different environments "are simply lacking" (Lewontin 1983: 277). But immediately after pointing out that the empirical data are simply lacking, Lewontin continued, incoherently, by insisting that "the little that is known shows clearly that the developmental responses of different genotypes to varying environments are non-linear and do not allow the simple ordering of genotypes along a one-dimensional scale of phenotype" (Lewontin 1983: 277). By the way, it seems that Lewontin's 1983 statement about the paucity of data on the significance of G–E

---

[6] I have assumed that the two genotypes (G_1 and G_2) were uniformly distributed across three focal environments.

interactions in natural populations is still true even after twenty years (cf. Boomsma & Martin 2002: 186).

Second, since the context in which Sterelny and Griffiths discuss the operation of genes and environments is human behavior genetics, their readiness to follow Lewontin in extrapolating so quickly the results from a couple of studies of fruit flies to people is quite surprising. It is worth recalling that, in comparison, sociobiological generalizations from animal research to humans had a much more solid empirical foundation but they were, as we know, condemned by most philosophers as unbearably crude and sometimes even as pseudo-scientific. Worse still, in the case of human behavior genetics there is actually no need to use a roundabout route through animal research. There is already a lot of direct empirical evidence (not mentioned by Sterelny and Griffiths at all) showing that in the human domain, G–E nonadditivity is extremely difficult to find, despite much research effort in that direction.

As if this were not bad enough, Griffiths goes further and asks: "Why do so many intelligent scientists appear to ignore facts that are well known to them, such as the likely nonadditive interaction of genotypes and environment?" (Griffiths 2002a). He thereby suggests that behavior geneticists behave very irrationally (they ignore facts that are well known to them). But why should they be so unreasonable? And, more importantly, where is the empirical evidence that they "ignore facts" (non-additivity) in the first place? Contrary to what Griffiths says, Lewontin actually does not provide that evidence. Nor, oddly enough, does Griffiths himself. He just issues a charge of massive irrationality but gives no supporting references, no data, basically no reason except Lewontin's authority. This best illustrates the mesmerizing influence that Lewontin has had on philosophers of science.

I am sure that in any other context it would be unimaginable that philosophers of science form a judgment about a highly contested scientific issue by trusting completely the word of one participant in the debate, even to the point of showing total disregard for other literature sources. In discussions about heritability, however, it has happened for some reason that the philosophical consensus is largely built on the basis of *ipse dixit*.

It would be a serious omission in this context not to discuss a famous study of rats conducted by Cooper and Zubek (Cooper & Zubek 1958), which is often mentioned in discussions about G–E interactions. Although this study is not mentioned in Lewontin's paper about the analysis

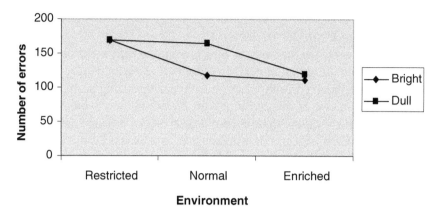

Figure 2.6 G–E interaction in rats. From Cooper and Zubek (1958).

of variance, it holds pride of place among empirical arguments for non-additivity, and has been called "the most widely cited report of heredity–environment interaction in psychology" (Wahlsten 1990a: 116), and "something of a classic in the nature–nurture controversy" (Platt & Stanislow 1988: 257). Many interactionists strongly rely on this paper in disputing additivity (Wahlsten 1995: 247; Wahlsten & Gottlieb 1997: 173–176; Kaplan 2000: 29–31, 62–63; Bateson 2001a: 151; Bateson 2001b: 565; Meaney 2001: 56–57; Pigliucci 2001: 259–260; Gottlieb 2003a: 344–345).

Two genetically different strains of rats (maze-bright and maze-dull) were tested in three different environments (restricted, normal and enriched). Figure 2.6 shows the mean number of errors these animals made under different conditions. In the normal environment there is a clear effect of genotype, but in the two other environments there is no significant phenotypic difference between the two strains. The norms of reaction of the two genotypes not being parallel indicates the presence of G–E interaction.

There are several reasons why one should be skeptical about the wide-ranging implications sometimes drawn from that experiment.

(a) Since it is just one study, any generalization from it must obviously be very shaky. As the study has never been replicated one should not rush to draw far-reaching conclusions from it. In fact, Henderson's experiments from the 1970s (see Plomin 1986: 98–99) involving thousands of mice revealed little statistical interaction, which shows that the empirical evidence here does not speak with one voice.

(b) There is no good reason to expect that the results of the Cooper and Zubek experiment on rats would immediately carry over into the area of human behavior genetics, although this kind of overhasty extrapolation has been defended (see Alland 2002: 96).

(c) As the rats belonged to two inbred strains it becomes even more questionable whether the results would be similar in normal organisms that are typically hybrid. (The point is made by Bouchard 1997: 143; Bouchard & Loehlin 2001: 261.)

(d) The observed interaction is not of a radical (non-ordinal) type, i.e., the norms of reaction do *not* cross. But it is non-ordinal interactions that most strongly undermine the attribution of main effects, and consequently the causal import of heritability claims as well. It is interesting to note here that Lewontin, with his insistence on the pervasiveness of non-ordinal interactions, seriously misrepresents Cooper and Zubek's study (although he doesn't mention the authors by name):

> strains of rats can be selected for better or poorer ability to find their way through a maze, and these strains of rats pass on their differential ability to run the maze to their offspring, so they are certainly genetically different in this respect. But if exactly the same strains of rats are given a different task, or if the conditions of learning are changed, *the bright rats turn out to be dull and the dull rats turn out to be bright.* (Lewontin 1993: 28 – italics added; cf. Lewontin 1977: 11)

The italicized phrase seems to suggest that the two genetically different strains of rats switch their positions on the bright–dull scale, but this actually never happened, as can be seen in Figure 2.6.

(e) Cooper and Zubek themselves warned about methodological limitations of their experimental set-up. They suggested that the two strains of rats might have actually differed in their *real* learning ability even in those situations where their performance was indistinguishable. It may have been, they said, that the ceiling of the test was simply too low to differentiate the animals (Cooper & Zubek 1958: 162). Moreover, they mentioned that something similar really happened with some tests of human intelligence "on which adults of varying ability may achieve similar I.Q. scores although more difficult tests reveal clear differences between them" (Cooper & Zubek 1958: 162). So there is a kind of vicious circle here. While critics of human behavior genetics often use Cooper and Zubek's experiment to raise methodological objections against research on IQ, unbeknownst to these critics, it is precisely the research on IQ that led the authors of that study to warn the reader that the results of their experiment should not be overinterpreted.

Let us leave Cooper and Zubek's rats rest in peace, and let us finally address the central question: how much interaction is there really? What does the empirical evidence say? Is the relation between genes and environments "typically non-additive," as Lewontin insisted, and as philosophers of science have usually been happy to repeat after him? It is actually amazing that the philosophical consensus could have been preserved for such a long time despite the fact that this view was under constant attack from people doing empirical research in the relevant fields. A few examples will illustrate the skepticism about massive interactions:

> Feldman and Lewontin should at least make explicit that their proposition is *purely a speculation* about as yet untested environmental conditions and that what data we do have concerning normal people moved around within common environmental situations show *their norm of reaction for IQ to be in fact rather flat.* (Havender 1976: 609 – italics added)

> nonadditive interactions rarely account for a significant portion of variance. (Plomin et al. 1988: 228–229)

> the simpler additive model in most cases comes close to fitting the expectancies. (Cattell 1982: 66)

> there is very little empirical support for [the] existence [of genotype–environment interactions] in the behavioral domain. (McGue 1989: 507)

> the data from the twins reveal no interaction (in the technical sense) of heredity and environment. (Herrnstein 1973: 180)

> since armchair examples of significant interactions in the absence of an additive effect are pathological and have never been demonstrated in real populations, we need not be unduly concerned about interaction effects. The investigator with a different view should publish any worthwhile results he may obtain. (Rao et al. 1974: 357)

> The Colorado Adoption Project analyses lead to the conclusion that genotype–environment interactions in infancy, if they exist at all, do not account for much variance. For this reason, Plomin and DeFries . . . proposed the following principle concerning the development of individual

differences in infancy: "Genetic and environmental influences of infant development coact in an additive manner." (Plomin 1986: 106–107)

Interactions are known to occur, but they are far from invariable occurrences. Under many, if not most, circumstances, effects are additive and, moreover, people tend to show broadly similar responses to the same environmental stimuli. (Rutter quoted in Plomin et al. 1988: 228)

Thus, as they stand, these results support the conclusion that genetic and environmental influences on individual differences in development in infancy and early childhood coact primarily in an additive manner. (Plomin et al. 1988: 240, cf. 250)

There is no conspiracy against interaction: If an interactive model could be shown to fit the data better than the traditional model, researchers would be quick to use it. (Plomin 1990: 144)

[The studies of interactions in nonhuman species have established that] although genetic control of sensitivity to the environment is widespread, the contribution of G × E to the overall population variance is typically smaller than the main effects of G and E even in controlled experiments using extreme environments. (Rutter & Silberg 2002: 465)

One aspect of genetic influence that environmental researchers seem to support is that of genotype–environment interaction. I agree that intuitively it seems as if these interactions must exist; however, genotype–environment interactions are non-existent in human literature. (Thompson 1996: 181)

on the whole it seems unlikely that disordinal interactions will be common. Instead I believe that "good" environments will be similar for all individual members of a species with a similar history of natural selection. Similar environments will not ensure equal performance; "better" environments may help some to achieve above the average. Rarely will a "bad" environment be good for any individual. G–E interactions are most likely to reflect quantitative variation in threshold of stimulation, persistence in problem solving, and slope of learning curves. (Fuller 1972: 473)

If variation in man has any similarity to variation in other organisms we would conclude that a trait was atypical if more than about 20 per cent of the measured variation could be attributed to G × E [interaction]. (Eaves et al. 1977: 3)

there is little or no formal evidence that personality and intelligence are determined by genetic × environmental interactions. (Brody & Crowley 1995: 66)

> Attempts to identify G × E effects for personality ... and general cognitive
> ability ... did not yield significant findings. (McGue & Bouchard 1998:
> 17)

> The most plausible explanation ... for the dearth of genotype–environment
> interactions in human behavior-genetics research on intelligence ... may
> be ... that genotype–environment interaction may not be important for
> individual differences in intelligence in most populations, given the ranges
> of intelligence, genotypes, and environments represented in these popu-
> lations. (Waldman 1997: 558, the word order slightly changed, for ease of
> citation)

Even the research on Drosophila, on which Lewontin drew so heavily
in his attack on additivity, sometimes points in the opposite direction and
fails to uncover massive interactions. For example, as Plomin (Plomin
1990: 144) reported, although some G × E interactions were uncovered
in a 1978 study of twelve Drosophila strains reared under twenty different
environmental conditions, the largest effect accounted only for 2 percent
of the total variance.

Speaking about attempts to determine empirically the scope of statistical interaction, it should be emphasized that, indeed, it is more difficult to
discover interaction effects than main effects. In statistical terminology,
the power of the test for interactions is lower than the power of the test
for main effects.

Radical and unjustified conclusions are sometimes drawn from this
asymmetry. Douglas Wahlsten, for instance, argued that inferring additivity in human behavior genetics is necessarily fallacious precisely because
the ANOVA non-experimental research design is especially insensitive
to the presence of interactions. To see the mistake in Wahlsten's argument, let us see what is right and what is wrong in his claim. Wahlsten
is right that if a given sample can detect only main effects of a given
size (say, those that account for more than 5 percent of the entire variance), then because of the asymmetry this sample will fail to detect some
interaction effects of that size (if they exist). But Wahlsten is wrong that
there is some intrinsic impossibility of capturing interactions in that way.
Put differently, more evidence is needed to accept the null hypothesis
about interactions than about main effects, but this certainly doesn't mean
that evidence will *never* be sufficiently strong to refute the existence of
interactions.

Although for a given effect size, the detectability of interactions will lag
behind that of main effects, it is still true that for a statistical interaction of

any magnitude there will always be a sample that is large enough to test for its presence. Also, the stronger an interaction, the easier to confirm it empirically.

Therefore, negative results in search for non-additivity support the hypothesis that there are no *massive* interactions (accounting for a high percentage of total variance), but they still leave room for the existence of weaker interaction effects. As Douglas Detterman said: "Wahlsten *must* agree that H × E interactions do not account for a large portion of the total phenotypic variance. If H × E interactions accounted for larger portions of the variance then even statistical techniques with lower power would be able to detect them" (Detterman 1990: 132).

This point is important because it is only in the presence of *strong* G–E interactions that it becomes difficult to put a useful causal interpretation on heritability coefficients. With a small contribution of interaction variance, however, this kind of problem does not arise.

Another argument against partitioning the variance into main effects is the non-experimental nature of research in human behavior genetics. Here the claim is that since for obvious moral reasons we cannot assign people with different genotypes to different environmental treatments, we will never be in the position to disentangle causal contributions of these two different types of causes: "In man, measurements of reaction norms for complex traits are *impossible* because the same genotype cannot be tested in a variety of environments" (Lewontin 1976a: 190 – italics added).

But this is again too quick. There are a number of ways to get around this difficulty and acquire causal knowledge *without* using experimental methods: comparing sums and differences of twins' phenotypic values (Jinks & Fulker 1970), adoption design (Plomin et al. 2001: 315–316), fitting models with and without an interaction term and seeing which one better conforms to available empirical data, etc.

In an important article, McClelland and Judd (1993) showed how samples with different preselected structures have different efficiencies for detecting interactions, and how researchers can maximize their chances of finding interactions (if they exist) by choosing samples that are specially conducive to this task.

The base square in Figure 2.7 represents 5 × 5 combinations of two variables, each of which takes five values: −1, −0.5, 0, 0.5, and 1. The heights of the columns show how many observations come from a

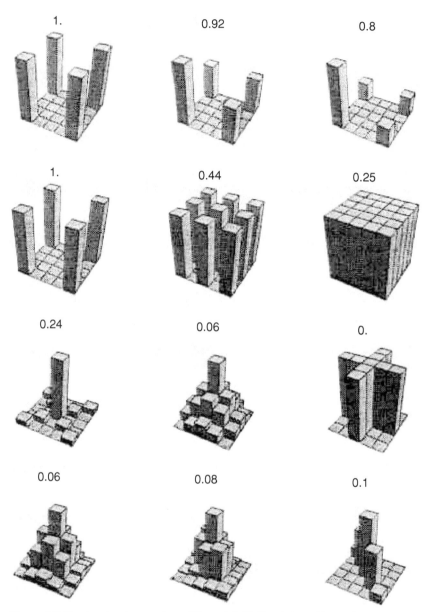

Figure 2.7 Testing for interactions (From McClelland & Judd 1993). Copyright © by the American Psychological Association. Reprinted with permission.

particular combination. We see that the optimal design for detecting first-order interactions is sampling at the extremes (corners). The statistical power for finding interactions decreases from the optimum (1.), and the decline in sensitivity depends on how a given sample deviates from the optimal design. Note that even when a full one-half of observations comes from one of the corners, the efficiency of detecting interactions is not significantly decreased, as long as the remaining observations are taken at other corners (the third case in the first row). This shows that oversampling at extremes is a good strategy in searching for genotype–environment interactions. It also shows that even without using an experimental approach there are ways to make research more interaction-sensitive.

It is interesting that despite their frequent invoking of empirical evidence, "interactionist" critics of behavior genetics actually tend to rely on aprioristic arguments in their defense of non-additivity. Here is a typical reasoning to that effect:

> because cognitive skills develop through complex and as yet little understood interactive processes involving many kinds of genetic and environmental factors, it would be unduly optimistic to expect such skills – even if they could be adequately measured – to fall into the rather narrow category of traits for which heritability is a meaningful concept. (Layzer 1972b: 423)

Now even if *at some level* developmental processes are highly interactive, context-sensitive, and extremely complex, it is a fallacy to conclude that *just because* of this "messiness" of developmental processes the main effects of heredity and environment are unlikely to be found. To use an analogy, the emergence of lung cancer must be a tremendously complicated process dependent on the combined action of many factors and coincidences, but despite all these unfathomable causal convolutions and elusive contingencies, we have very good reasons to believe that there is a quantitatively specifiable main effect of smoking on cancer. Layzer's argument (defended by many other authors) that complexities of developmental processes preclude the possibility of partitioning the phenotypic variation into genetic and environmental components seems to be the result of confusing different levels of analysis. This conflation is clearly explained by H. H. Goldsmith:

> Individual differences are the stuff of behavioral genetics, and classic behavioral genetics inferences are confined to genetic and environmental effects

73

on phenotypic variance, not genes and environments per se. There is no contradiction in analyzing individual differences by linear regression of outcome on sources of variation, even when the individual differences result from highly contingent developmental processes operating in the life of individuals. In fact, psychologists frequently do analogous exercises. For example, both earlier IQ and quality of schooling might predict later academic achievement of children in a linear fashion. Computing the relevant regression and interpreting the partial regression coefficients is a legitimate and potentially useful exercise even though the actual learning experiences of the children were highly contingent, interpersonal, and context bound. Analyzing the nature of the contingencies and contextual influences is simply a different task. (Goldsmith 1993: 329)

Many developmentalist criticisms of heritability (advocated by Wahlsten, Gottlieb, Oyama, etc.) would be more relevant if they showed that they appreciated the difference between the two tasks and that they did not try to prove, hastily and fallaciously, that the separation of causal contributions is impossible *only because* these causes work together in development in a very complicated way.

When the authors of the DST manifesto mention the question whether intelligence is 50 percent or 70 percent genetic (the natural interpretation of "genetic" in this context being "heritable"), and when they say that "DST rejects the attempts to partition causal responsibility for the formation of organisms into additive components" (Oyama et al. 2001: 1), they seem to forget that additivity is ultimately an empirical issue and that it cannot be decided by *a priori* speculation. Their descent into the worst tradition of *Naturphilosophie* is obvious in the following statement: "Whenever a number of causal factors interact to produce an outcome, we should expect that the effect of changing one factor will depend on what is happening to the others" (Oyama et al. 2001: 3). How do they know? On what basis do they state that we should expect non-additivity *whenever* a number of causal factors interact to produce an outcome? This is a dogmatic and arbitrary claim. Moreover, we do not have to wait to see whether, so to speak, the planet Ceres will be discovered. They are already proved wrong now because those contexts in which additivity or near-additivity reigns represent a straightforward refutation of their "whenever" statement.

Notice that, using the terms from the beginning of this chapter, Oyama et al. basically infer interaction$_s$ (statistical interaction) from interaction$_c$ (commonsense interaction). Their conclusion (what "we should expect") is that "the effect of changing one factor will depend on what is happening to the others." But this is actually how interaction$_s$ is defined. Therefore, when they speak about interaction in the premise ("Whenever a number of causal factors *interact* to produce an outcome") they cannot mean interaction$_s$, because this would make their inference tautological. Surely, they don't want to tell us that whenever there is interaction$_s$ we should expect interaction$_s$. The most charitable interpretation is that by "interaction" in the premise they mean interaction$_c$, and then their inference is fallacious rather than tautological.

The idea that the effect of changing one factor *always* depends on what is happening to other factors is an empirical claim, and a demonstrably false one at that. A dramatic refutation is what happens with genetically modified organisms, where particular genes often continue to have the same effect even after being transplanted across species and placed into a drastically different genetic and environmental context:

> For example, the soil bacterium *Bacillus thuringiensis* produces a protein that is toxic to caterpillars. Scientists have placed a gene from these bacteria into plants, such as cotton, that are extremely susceptible to insect damage. When a caterpillar takes a bite of one of these modified plants, it consumes some of the toxic protein and dies. (Crawford 2003: 856)

2.5 LOCALITY

If genetic and environmental effects happen to be additive in a given population, can it be inferred from this that additivity will then probably exist also in similar populations or in the same population under slightly changed conditions? Can a heritability value obtained in one population be extrapolated to some other situation? The answer is often a resounding "No," on the grounds that heritability is a population statistic, and that it applies *only* to the population for which it was measured and under the exact circumstances in which it was measured. And indeed, if the claim of massive non-additivity and pervasiveness of statistical interaction is correct, then a high heritability value measured in a given range of environments will rightly be regarded as an "atypical" case of additivity and it will be expected to revert to a tangle of interactions as soon as the population parameters are even slightly altered. As Lewontin said: "the usual outcome of an analysis of variance in a particular population in a restricted range of environments is to underestimate severely the amount of interaction between the factors that occur over the whole range" (Lewontin 1976a: 190).

Although, strictly speaking, it follows from Lewontin's statement that one is actually entitled to infer that heritability will be *lower* whenever the range of environments is broadened (because the "usual outcome" is that the amount of interaction is "underestimated" in a restricted range), the prevalent view is that *no* inference is permissible to new situations.

It is useful at this point to distinguish between two statements about heritability and its generalizability: (1) a heritability value obtained in a given population is not *automatically* generalizable to *all* other populations; (2) a heritability value obtained in a given population can *never* be extrapolated to *any* other population. Claim (1) is trivially true and hence uninteresting, while claim (2) is interesting but false.

(1) When Patrick Bateson says that "the heritability of any given characteristic is not a fixed and absolute quantity – tempted though many scientists have been to believe otherwise" (Bateson 2001a: 150), what immediately catches the eye is that he gives no references, no names. Who are these "many scientists" who believe that the heritability of any given characteristic is "a fixed and absolute quantity"? My bet is that these scientists do not exist, particularly not in behavior genetics. Competent people in the field simply do not make such a mistake. Similarly, when Ned Block (Block 1995: 108) maintains that there is "the temptation (exhibited in *The Bell Curve*) to think of the heritability of IQ as a constant (like the speed of light)," there is no quotation to support that charge. The reason for this absence is again non-existence. To make things worse, Block contradicts himself on the same page when, in criticizing Charles Murray's statement from a CNN interview, he writes: "*The Bell Curve* itself does not make these crude mistakes. Herrnstein, the late co-author, was a professional on these topics." How can Herrnstein be a professional on these topics if he thought that heritability is a constant, like the speed of light? The implausibility of Block's accusation[7] is manifested in the fact that already in his classic 1969 paper Jensen stated in passing, as a matter of obvious truth: "From what has already been said about heritability, it must be clear that it is not a constant like π and the speed of light" (Jensen 1969b: 43). Herrnstein and Murray also explicitly warn "nonspecialists"

---

[7] Spencer and Harpalani (2004: 57), who quote Block's paper, also state that in *The Bell Curve*, heritability is misrepresented as "a static, unalterable entity."

that heritability is *not* a constant: "The heritability of a trait may change when the conditions producing variation change" (Herrnstein & Murray 1994: 106).

(2) After we reject the idea (held by no one) that heritability measured in one population immediately applies to *all* other populations, how about inferring the magnitude of heritability in a population from the one measured in some other population? We know that this kind of inference is not always legitimate. But is it ever legitimate? Notice that I am talking about inductive inference here, not a deductive relationship. So, when Alland says that "a measure of heritability in one population, let us say 0.80, is *no guarantee* that the heritability will be the same (for the same trait) in another population in another environment" (Alland 2002: 91; italics added), this is easily conceded. No guarantee, of course, but is there at least a high probability that a measure of heritability will carry over from one population to another, *in some cases*? The negative answer seems to be given in the following passages:

the [heritability] figures are population specific; that is, they apply *only* to the particular *samples* studied *at a particular time*. (Rutter 2002: 2 – italics added)

Heritability estimates are necessarily specific to the particular time and to the sample from which the measures were collected. Heritability estimates describe only what is the case now and have no implications for what could, or would, happen if circumstances changed. (Rutter 1997: 391)

But heritability could be 1.0 in one condition and near zero in another, for heritability is condition dependent, as already discussed. The most that can be said from a high value of heritability is that phenotype determination is highly influenced by genotype *in the conditions observed*. (West-Eberhard 2003: 102–103)

it is *almost impossible* to extrapolate the heritability of a character for one population from information about another. (Daniels et al. 1997: 54 – italics added)

heritability applies *only* to the set of environments in which the determination was made and cannot be extrapolated to other distributions of environments. (Lewontin 1976b: 9)

whatever [heritability] tells us is relative to that population and the existing sources of variation it tells us *nothing whatever* about populations that differ from the initial one in genetic and environmental variation. (Block & Dworkin 1976b: 486–487 – italics added)

Heritability estimates are relevant only for the specific environment in which they are measured. (Nelkin & Andrews 1996: 13)

Interpreted literally, Rutter's two passages make heritability figures totally useless because they carry information only about *samples* at a given time (not even about populations at a given time). West-Eberhard also robs heritability estimates of any interesting content because they can never go beyond what is directly observed. Daniels and his colleagues use an odd phrase "almost impossible" to describe the prospects for a population-to-population extrapolation of heritability.

Since the main reason for disputing the generalizability of heritability is that it is a population statistic, the argument is clearly quite general (it is not restricted to behavior genetics). But this is precisely what makes it unconvincing. Can we really agree that statistical information obtained with ANOVA methodology applies *only* to samples that were examined in the initial phase of research, and that this information is *totally irrelevant* for any context that transcends the narrow evidential base? Indeed, to raise this question is to answer it negatively.

The more similar a new context is to the examined one, the stronger the expectation that the inference from one case to the other will not be mistaken. For example, if there was a ratio of 3:1 between the incidence of lung cancer and stomach cancer in a given country in 1990, then one would not be surprised if that particular ratio was of the same order of magnitude in 1980 or in 2000. Surely, if the only thing one knew was that in 2000 there were approximately three times more cases of one kind of cancer than the other, one would have more reason to believe that the 1990 ratio was preserved than that it was reversed. The inference would be fallible, no doubt, but the fewer changes in these ten years, the more secure the extrapolation. Moreover, a cautious and appropriately qualified generalization of that ratio to *other* populations that did not differ much from the country in question cannot be *a priori* dismissed as illogical.

Yes, a heritability coefficient refers to a particular trait in a particular population, but "this does *not* mean that if one finds a given trait to be highly heritable in one human population, he should be astonished to find it also highly heritable in another" (Loehlin et al. 1975: 81 – italics added). One should actually expect similar heritabilities in similar populations. And conversely, "the greater [the differences between the populations] are, the riskier the generalizations will be" (Loehlin et al. 1975: 81).

If the "locality extremists" looked into the standard introduction to quantitative genetics they would see that heritability estimates typically go beyond the information given and can readily be extrapolated to other (similar) cases:

> whenever a value is stated for the heritability of a given character it must be understood to refer to a particular population under particular conditions. Values found in other populations under other circumstances will be more or less the same according to whether the structure of the population and the environmental conditions are more or less alike. (Falconer 1989: 164)

The question of generalizability of heritability estimates is ultimately an empirical issue (cf. Bouchard & Loehlin 2001: 247). For this reason, blanket condemnations of any such inference are misconceived. Ned Block (1995: 108) claims: "There is no reason to expect the heritability of IQ in India to be close to the heritability of IQ in Korea." How does he know? The troubling aspect of Block's resolute statement is that it is in fact directly contradicted by relevant empirical research:

> The question of generalizing [the components of variance] to other samples and other times can only be answered empirically (the evidence with respect to cognitive abilities suggests *considerable generalizability*). (Plomin & DeFries 1976: 11 – italics added)

Block also ignores one of the most important contributions to that topic, which was published six years before his paper. In an article "The Generalizability of Genetic Estimates" (in the leading psychological journal *Personality and Individual Differences*), J. Philippe Rushton concluded from research on Japanese-American, European-American, Korean, British, Australian, and Canadian samples that "from the data reported here, estimates of genetic influence appear to have a greater robustness across populations, languages, time periods and measurement specifics than has been considered to date" (Rushton 1989: 988). Needless to say, Rushton's conclusion may be disputed, but for someone to discuss the issue without even mentioning this important contribution to the literature is not the best example of responsible scholarship. By the way, later research seems to have confirmed Rushton's view: "Although heritability could differ in different cultures, moderate heritability of *g* has been found, not only in twin studies in North American and western European countries, but also in Moscow, former East Germany, rural India, urban India, and Japan" (Plomin 2002: 213).

David Moore radicalizes Block's position by saying that the generalization of heritability is unjustified not just across countries but across state borders in the United States as well: "If alcoholism is heritable among Iowans, it need not be the case that it is heritable among Ohioans. (I am not simply being recalcitrant here: heritability estimates calculated for one population *do not apply* to another population)" (Moore 2001: 47). Moore's position approaches the limit, i.e., the idea that any heritability value is valid only within the sample on which it was calculated.

Critics of heritability seem to be unaware that because of the so-called "causal democracy" (the symmetrical status of genetic and environmental factors), the non-generalizability of heritability implies the non-generalizability of environmental influences as well. Therefore, it would follow from Moore's pessimism about state-to-state inferences that if a new teaching strategy had good effects in schools in Ohio there would be no reason whatsoever to expect that the strategy would work in Iowa. This consequence is absurd, but it should not be surprising that it can be derived from the criticism of heritability, because that criticism is not based on specifics of genetic causation. On the contrary, it has a very general import and is bound to have broad implications since it applies to any ANOVA causal inference, independently of the subject matter. This is why, when fully understood, it looks less like an argument about behavior genetics and more like a full-fledged inductive skepticism.

Interestingly, the non-experimental nature of research in human behavior genetics, which is frequently regarded as its serious methodological deficiency (Michel & Moore 1995: 209), may actually work in its favor in this context. Not dealing with artificial, laboratory conditions but with real-life situations, its conclusions obtained on given samples are easier to extrapolate to other similar populations: "Human behavioral genetic researchers have the mixed blessing of working with naturally occurring genetic and environmental variation. The cost is a loss of experimental control; the benefit is an increased likelihood that the result of the research will generalize" (Plomin et al. 1988: 231; cf. Plomin 1986: 99; Plomin & Hershberger 1991: 33). Causal claims in human behavior genetics are more difficult to confirm, but once they gain plausibility they have more ecological validity than studies using clones-in-cages design.

2.6 CAUSAL IRRELEVANCE

Even if heritability estimates are generalizable within a given range (of environments and genotypes), it seems very likely that, if pushed far enough, these generalizations will eventually break down. This fact is used to argue that heritability claims cannot give us knowledge of causal relations between genes and phenotypes. Here is Lewontin again:

> [the result of the ANOVA approach] has a historical (i.e., spatiotemporal) limitation and is not in general a statement about *functional* relations. (Lewontin 1976a: 183)

> the analysis of variance will give a *completely erroneous picture of the causative relations* between genotype, environment, and phenotype because the particular distribution of genotypes and environments in a given population at a given time picks out relations from the array of reaction norms that are *necessarily atypical of the entire spectrum of causative relations*. (Lewontin 1976a: 188 – italics added)

> In view of the terrible mischief that has been done by confusing the *spatiotemporally local analysis of variance* with *the global analysis of causes*, I suggest that we stop the endless search for better methods of estimating useless quantities. (Lewontin 1976a: 192 – italics added)

Why is ANOVA an inappropriate instrument in the search for causes? In Lewontin's picture, since ANOVA has a *historical* (*spatiotemporal*) limitation and is necessarily restricted to *local, particular* distributions of the relevant variables, it cannot deliver the *global* analysis of causes, or knowledge of *functional* relations and of *the entire spectrum of causative relations*.[8] But how can we hope to get knowledge of these functional or global causal relations? Well, they are "embodied in the norm of reaction" (Lewontin 1976a: 189), but unfortunately "in man, measurements of reaction norms for complex traits are *impossible* because the same genotype cannot be tested in a variety of environments" (1976a: 190 – italics added). It follows that it is impossible to know how genes and environments cause human phenotypic differences.

Lewontin is right that *if* causal knowledge in behavior genetics were contingent on knowing the full norms of reaction then such knowledge

---

[8] The DST manifesto also quotes Lewontin as showing that "heritability estimates are not measures of global causal importance" (Oyama et al. 2001: 3; cf. Oyama 1992: 221; Gray 2001: 187).

would be absolutely outside of our reach. Norms of reaction cannot be known in their entirety:

> Complete knowledge of a norm of reaction would require placing the carriers of a given genotype in *all possible* environments and observing the phenotypes that develop . . . The performance of a genotype *cannot be tested in all possible environments*, because the latter are infinitely variable. (Dobzhansky 1955: 75 – italics added)

> To test the reactions of a given gene constellation (or a cluster of kindred ones) in all environments is obviously impossible. For example, how could one discover the highest stature that I could have reached in some very propitious environment, or the lowest one at which I could have stopped growing and still remained alive? (Dobzhansky 1973: 8)

> As a rule, it would be difficult or even *impossible* to denote the whole range of the norm of reaction (NOR) of a characteristic, and we must be content with a partial [norm of reaction]. (Falk 2001: 121 – italics added)

> Ideally one would like to know the reaction of every genotype in every environment. Given the practically infinite variety of both environments and genotypes, this is *clearly impossible*. (Bodmer & Cavalli-Sforza 1970: 24 – italics added)

Evidently, the dubious part of Lewontin's position is his making the complete knowledge of norms of reaction a *necessary condition* for understanding causal influences of genes and environments on the phenotype. First of all, why should we accept that the search for causal explanations always has to have as its goal the "*global* analysis of causes," or knowledge of "*functional*" relations across "*the entire spectrum* of causative relations"? In many etiological contexts researchers settle for a less ambitious and less grand task of finding causal relations that obtain only under contingent and antecedently specified conditions. In biology, in particular, a maximalist epistemic aim of looking for *global* causal connections sounds distinctly unrealistic and utopian.

For instance, saying that a chromosomal abnormality associated with Down's syndrome *causes* a significantly lower IQ does *not* commit us to the claim that, *globally*, a norm of reaction of that genotype will *always* (in all possible environments) be below the norm of reaction of a normal genotype. If the distance between the two norms of reaction has been consistently preserved over a number of environmental regimes that had been tried (including those introduced in the hope of eliminating the difference in question), then we are perfectly entitled to speak about causality, *even if we allow that there might be some (presently inaccessible) environments where the difference disappears.* A good slogan for many biological contexts would be: observe locally, think causally (and forget "globally").

Imagine that we discover a linear causal relation between human height and intake of calories, according to which, in the developmental period after birth, a yearly intake of $n \times X$ calories results in the growth of n inches. Suppose now that someone disputes the causal connection between the two variables by pointing out that the functional relation is not global, i.e., that it breaks down when n becomes very small or very big. Babies exposed to an extremely poor diet (small n) do not grow to become 2–3 inches tall in adulthood. Likewise, babies absorbing drastically large amounts of food do not grow to "predicted" heights of 20–25 feet. (I am indebted for this analogy to Turkheimer et al. 1995: 147.)

This criticism would rightly be dismissed as ridiculous because, if cogent, it would undermine practically all causal explanations except perhaps those in fundamental physics. In biology, as in most other explanatory contexts, we know in advance that any causal relation will break down if the regime of the relevant parameters is pushed far enough beyond a given range.

The weaknesses of this "possible-worlds" approach to the heredity–environment controversy are best recognized in Kitcher's attempt at reconstruction of the nature–nurture debate. He starts by supposing that each camp misrepresents the opposite standpoint by understanding it to claim that only genes or only environment determine phenotype: too facile a victory is won by attacking these men of straw. Kitcher then argues that there must be a point of real and rational disagreement between the two sides, even after they acknowledge the "interactionist truism" – the fact that *both* genes *and* environment determine phenotype. His proposed solution is to burden the "genetic determinists" with a claim about what happens in *all possible* environments (Kitcher 1985: 25–29; Kitcher 1987: 66). Accordingly, genetically determined phenotypic differences are identified as those that either cannot be eliminated or are too costly to eliminate, *for any possible environment.* Kitcher believes that this reinterpretation is fair to hereditarianism, in that it avoids making it either simplistic ("*only* genes determine phenotype") or unexciting ("genes have *some* influence on phenotype").

It is more than doubtful, though, that real hereditarians will be happy with such a division of roles. They would most probably regard playing the part assigned to them in the debate by Kitcher as equivalent to admitting defeat. For, the one who takes upon himself the Herculean task of proving that something holds literally *under all circumstances*[9] should be aware that he has no defense against Kitcher's quite correct claims about "the immensity of our ignorance about the environmental influences on human behavior" (Kitcher 1985: 29) and about "our [profound] ignorance of the shapes of the norms of reaction" (Kitcher 1987: 66). Human behavior geneticists are checkmated in two moves. First, they are told that they should try to discover human norms of reaction in their entirety, but then, second, they are told that it is *impossible* to trace out human norms of reaction for any trait (Lewontin 1977: 10). Chess, anyone?

But our criticism should try to strike at a deeper level. Let us therefore raise the following question: is the norm of reaction approach superior to other approaches in some important theoretical respect? That is, would we lose anything if we decided to abandon the search for complete norms of reaction,[10] and if we instead adhered to the precisely defined concept of statistical interaction? Robert Plomin argues that we would thereby only free ourselves from many vague connotations which the term "norm of reaction" drags in its wake:

> Indeed, the connotations of the phrase may be responsible for its popularity. Reaction range suggests the notion that, regardless of genotype, an individual can grow up to be a pauper or a prince depending on environmental circumstances. Although such plots make good fairy tales, all we can assess is environmental variance, genetic variance, and variance due to genotype–environment interaction. The term *reaction range* adds nothing to these concepts. (Plomin 1986: 95)

Plomin's "fairy tale" irony is here remarkably pertinent. For, the norm of reaction (as used by some authors) does indeed tend to divert our attention from the manageable empirical data over which we have some control, and directs it to the question of what might have been the case in some extremely different and not very clearly defined circumstances. Speculation about these remotely conceivable situations may then begin to dominate the picture so much that we witness the curious triumph of

---

[9] According to Kitcher (Kitcher 2001b: 399), the norm-of-reaction approach requires that "the *entire* space of nongenetic causal variables is covered" (italics added).

[10] Notably, when Schiff and Lewontin analyzed eleven genetics textbooks written by prominent scientists in the field (Schiff & Lewontin 1986: 188–189), none of them contained a reference to the norm of reaction. My search gave a similar result. Despite being so often criticized, the concept of heritability is extensively discussed in all textbooks, whereas the much advertised norm-of-reaction approach is usually not mentioned at all.

the *possible* over the *actual*. We should resist this and try to interpret the heredity–environment discussion so that it deals with our real world, not with uncontrollable "might-have-beens" and counterfactuals gone wild. But then we would not get very far with the "norm of reaction," understood in the aforementioned way, which only encourages flight into speculation. Rather, it would be advisable to aim at discovering the connections that hold non-globally, the only piece of knowledge that remains a realistic research goal.

The following quotations illustrate the two opposite views that might be called "possibilist" and "actualist" standpoints:

> We suspect that dissemination of the idea that genes and other developmental factors *may* interact with the environment in multifactorial, non-additive ways to produce outcomes would greatly improve the debate over the role of genetic factors in determining human behavior. It is probably fair to say that many people assume additivity when discussing gene–environment interactions simply because they overlook *the full range of the possible*. (Sterelny & Griffiths 1999: 16–17 – italics added)

> The scarcity of evidence for interaction does not mean that interaction is forever impossible, only that it is not accounting for much now. It is, in fact, entirely possible that science could uncover ways of raising people's I.Q.'s by special sorts of environments, tailor-made for them. A world in which each person enjoyed something approaching his optimal environment – let us assume a different environment for each – might register large interaction and little overall variation in I.Q. That is, however, not our world, and we have as yet hardly an inkling of how to get from here to there, or even of whether or not the way exists in any practical sense. (Herrnstein 1973: 180)

It is easy to agree with Sterelny and Griffiths that we should keep an open mind to the possibility of gene–environment interactions in new, unexplored domains of environmental variation. But it is harder to see how "the idea that genes . . . *may* interact with the environment . . . would greatly improve the debate over the role of genetic factors in determining human behavior." True, genes may interact with environments, but then again they may not. Surely, both possibilities should be taken seriously, yet perhaps not too seriously as long as they are just that, mere possibilities. "Overlooking the full range of the possible," which Sterelny and Griffiths condemn as a methodological shortcoming, is better regarded as the characteristic bias toward actuality that is the essential mark of empiricism.

85

To recapitulate, the result of our discussion is that Lewontin's argument has problems in each of its three steps. First, the claim of massive non-additivity is not supported by empirical evidence. Second, the claim of locality is false if interpreted as saying that any generalization of heritability is unjustified. And third, the claim of causal irrelevance fails because its explanatory pessimism mainly springs from describing the research goal of behavior geneticists in such maximalist and unrealistic terms that it sends the whole field on an epistemic "mission impossible."

2.7 THE SECOND LOOK AT INTERACTIONS

The use of ANOVA in behavior genetics (and elsewhere) requires great caution. Applying the technique uncritically can easily lead to inferences that distort the picture of actual causal connections. Therefore, it is perhaps good from time to time to be reminded (e.g., Wahlsten 1990b: 152) that a large sample size is necessary for the test of statistical interactions, that if the functional relationship of two factors is non-additive then it may be difficult to speak of separate causes of variation, that ANOVA is appropriate in some situations but not in others, and so forth. But pointing to all these possible pitfalls and limitations cannot support the strong claim that we should abandon the goal of partitioning variance among mutually exclusive causes and calculating heritability coefficients (Wahlsten 1990b: 109), or that the ANOVA approach is somehow irremediably and intrinsically defective as an instrument in the search for causes. An occasionally misused method is not a useless method.

It is sometimes suggested that the analysis of variance imposes a straitjacket of additivity on the world that is fundamentally characterized by non-additivity. This is also wrong. True, ANOVA gives useful results only if measured factors are largely additive, but if they happen not to be, no reasonable person would be inclined to continue pushing this approach. There is nothing in ANOVA itself that leads to its being applied in totally inappropriate contexts. As Ronald A. Fisher says: "[T]he main point is that you are under no obligation to analyze variance into its parts if it does not come apart easily, and its unwillingness to do so naturally indicates that one's line of approach is not very fruitful" (in Bennett 1983: 218).

Another objection is that, rather than starting with idealizations that abstract from the world's complexity, the research should already at the beginning face reality in all its glory and richness of intricate causal connections and interactions: "Causal interconnectedness in the world is not

DST's invention. We just deal with it differently, by including it in our formulations from the start, rather than bringing it in only when forced to do so, and then perhaps marginalizing it in some way" (Oyama 2000b: S344). Is this actually a good idea? To motivate skepticism about Oyama's proposal let me quote an opposite view defended by behavior geneticist John Loehlin in another discussion: "This is a bit like saying that when a train hits a bus, events are extremely complicated at the level of the molecules involved. Well, yes, they are, but is that the level at which to look at things first?" (Loehlin 2001: 169). Or, as William Thompson said: "In the initial stages of a genetic analysis it might be better to oversimplify than to overcomplicate" (Thompson 1975: 1126).

Critics of behavior genetics often point to the enormous complexity of developmental processes, as if this fact were unknown to psychologists who talk about heritability of different traits. Gilbert Gottlieb, for example, repeatedly argues that behavior genetics is "conceptually deficient," and he likes to warn those working in that field about complexity by giving a graph in which intricate causal relations between different levels of biological organization are displayed (Gottlieb 1995: 138; Gottlieb 2003b: 11; Gottlieb 2004: 95; Wahlsten & Gottlieb 1997: 183). Is this kind of reminder really necessary? I think not, because no researcher worth his salt ever thought that the partition of phenotypic variance into components *exhausts* all the relevant information about causality of trait development. The underlying complexity is fully recognized and taken for granted, but the claim is that even causally most convoluted situations still leave open the possibility of additivity of genetic and environmental influences.[11] That behavior geneticists are indeed perfectly aware of the causal labyrinth and interrelations between different levels of organismic functioning is best reflected in the fact that Gottlieb's diagram, which he constantly uses as a weapon of criticism of behavior genetics, is old news in the discipline he criticizes. We can find a strikingly similar graph as an illustration of biological complexity in a popular overview of research in behavior genetics that was published more than forty years ago (McClearn 1964: 167).

Searching for main effects in the nature–nurture debate certainly involves a simplification, but the main question is whether it is a fruitful simplification. If there is something fundamentally wrong with this

---

[11] "Fortunately, relationships at the population level may often be additive, or nearly so, over the ranges of genetic and environmental variation found in natural situations, even though the underlying biochemical and behavioral processes are highly complex and non-linear" (Loehlin et al. 1975: 79).

method, we should condemn it in other situations as well. Do we really want to do that?

Let me use an argument from analogy. Suppose that it is discovered that in criminal court cases white judges (and white judges only) sentence black defendants more harshly than white defendants. On the basis of this observation it could be argued that there is a causal tendency here, whereby one variable (the race of the judge) influences another variable (the severity of sentence for black defendants). This could be taken as strong prima facie evidence of racism in the legal system.

But imagine now that person X resists the conclusion by saying that the "development" of a judicial decision is an extremely complicated process, and that it is unbearably crude and naive to think that we can separate or measure the influences of two factors that interact to produce the sentence (the two factors being the race of the judge and the nature of the crime). X may also insist that the observed correlation between the race of the judge and the severity of sentence for blacks applies only to the sample from which the data were obtained, and that no justification exists for extrapolating this regularity from the observed sample to any other context (e.g., from Iowa to Ohio). Furthermore, now X can no longer be stopped in his Lewontinian outburst, the fact that the "norm of reaction" of white judges is consistently higher than the one of black judges (the white judges tending to react to black defendants with harsher sentences than black judges) is just a local phenomenon. It may well be, he continues, that there are situations (not yet explored) where the two norms of reaction cross, which all just shows that the available statistical data are insufficient to warrant any conclusion that the race of the judge is *causally influencing* the severity of sentence: the analysis of variance is different from the analysis of causes.

Wouldn't we rightly regard this train of thought as a disingenuous attempt to turn a blind eye to the strong evidence of judicial prejudice based on race? This analogy shows that by moving to other contexts of discussion we can recognize fully the radical nature of Lewontin's methodological purism. Following his advice would throw out many innocent and very healthy babies together with the alleged bathwater of hereditarianism.

3

Lost in correlations? Direct and indirect genetic causes

Unless you are willing simply to deny that causality is a meaningful concept then you will need some way of studying causal relationships when randomized experiments cannot be performed. Maintain your skepticism if you wish, but grant me the benefit of your doubt. A healthy skepticism while in a car dealership will keep you from buying a "lemon." An unhealthy skepticism might prevent you from obtaining a reliable means of transport.

Bill Shipley

The fact that organisms with genotype G_1 may have phenotype P_1 more often than organisms with genotype G_2 is not necessarily a good indication that a G-difference is *directly* responsible for a P-difference. It may well be that (for whatever reason) G_1 organisms simply find themselves more frequently in environment E_1 than G_2 organisms, and that it is E_1 which causally leads to phenotype P_1 (while E_2 produces P_2). This phenomenon is called "genotype–environment correlation." It has been extensively discussed in the behavior genetics literature, but it has also been widely used in methodological criticisms of heritability.

The expression "genotype–environment correlation" usually refers to the situations where two separate sources of phenotypic variance (genetic and environmental) happen to be correlated. Sometimes, however, it is interpreted more broadly to cover also the cases where there is a correlation between a genetic and environmental characteristic, even when the genetic characteristic is not *directly* influencing the phenotype. For example, if a difference between G_1 and G_2 is in itself irrelevant for the phenotype but if this genotypic difference is statistically correlated with a trait-relevant environmental difference, this situation would be regarded as the case of G–E correlation on the second interpretation (because there would be a correlation between a genetic characteristic and an environmental influence on phenotype), but it would not count as such on the first interpretation (because genetic and environmental differences would not both be separate and independent sources of phenotypic variance). I will use the broader connotation because the criticism of heritability is sometimes couched in these terms.

There are two main objections to heritability that are based on the existence of genotype–environment correlation: (a) conceptual and (b) methodological. Allegedly, the ever-present possibility of G–E correlation shows (a) that the notion "heritability" is logically odd because it does not keep genetic and environmental influence conceptually separated, or (b) that the genetic and environmental influences, although conceptually distinct, cannot be empirically disentangled from one another. These two strands of argument are not always clearly distinguished. For clarity and better focus I will discuss them separately. First, the conceptual tack.

3.1 THE CONCEPTUAL ROUTE: A PICKWICKIAN NOTION?

Does the fact that a given phenotypic trait is heritable entail that it is genetic (i.e., that the differences in that trait are due to genetic differences)? Many influential social scientists and philosophers (Christopher Jencks, Ned Block, Elliott Sober, Allan Gibbard, and others) give a negative answer to that question. They all support this answer by using an essentially same example, which originates from Jencks et al. 1972.

3.1.1 The redheads example

Imagine that red-haired children are for some reason singled out for abuse and are frequently hit on the head by parents and teachers. As a result they get a lower IQ on average than other children. Now, although it is in a certain sense true that in this situation having a gene for red hair leads to a lower IQ, it just doesn't sound right to say that the IQ deficit of red-haired children is genetic. For, the manifestly critical influence here is the environmental one (the abuse), which explains the deficit, and furthermore explains it *completely*. Jencks and his followers claim that according to the way "heritability" is used in behavior genetics, the IQ difference between red-haired and other children in this hypothetical example would be heritable and counted as genetic. Since this strongly conflicts with common sense (which regards that difference as 100 percent environmental) they conclude that heritability is a somewhat Pickwickian causal notion, and that for that reason, the heritability of a trait by no means implies that the trait is also genetic in the important sense of that word which we ordinarily use to divide causal responsibility between genes and environments. Here are several characteristic passages:

> If, for example, a nation refuses to send children with red hair to school, the genes that cause red hair can be said to lower reading scores . . . Attributing redheads' illiteracy to their genes would probably strike most readers as absurd under these circumstances. Yet that is precisely what traditional methods of estimating heritability do. If an individual's genotype affects his environment, for whatever rational or irrational reason, and if this in turn affects his cognitive development, conventional methods of estimating heritability attribute the entire effect to genes and none to environment. (Jencks et al. 1972: 66–67)

> It is a by-product of the methodology for measuring heritability to adopt a tacit convention that *genes are taken to dominate environment* . . . If there is a genetic difference in the causal chains that lead to different character-istics, the difference counts as genetically caused even if the environmental differences are just as important. (Block 1995: 117)

> But the methods used to assess the heritability of IQ automatically count variance produced by genetic variation as genetically caused variance even if it is also environmentally caused. (Block & Dworkin 1976a: 480)

> imagine [that] blue-eyed children are fed to lions, but some of them survive, maimed. If eye color is inherited and this grim ritual is the predominant cause of anyone lacking a leg in that population, then non-two-leggedness in that population has substantial heritability. (Gibbard 2001: 169)

> If blacks are badly treated because of their skin color, and their skin color is genetic, then the lower IQ will be assigned to genes, not to environment. (Sober 2001: 74; cf. Sober 2000: 366)

(See also Levins in Callebaut 1993: 249–250; Jencks 1980: 730; Garfinkel 1981: 119–120; Jencks 1992: 106; Taylor 2001: 179; Wachbroit 2001: 39; Wasserman 2004: 25; Billings et al. 1992: 230; Moore 2001: 46; Parens 2004: S12.)

Summarizing his views on the heredity–environment controversy, Jencks (1980: 731) gives a two-by-two table, containing a taxonomy of various kinds of phenotypic effects. For our present purposes it is worth noting that differences resulting from sexual discrimination are classified there as being *genetic*! These differences are of course usually regarded as being purely environmental in origin. The surprising outcome of labeling them as genetic directly follows from Jencks's interpretation of heritability, according to which any variation ultimately caused by genetic variation is heritable (and hence genetic).

Clearly, sexual discrimination is initially triggered by a genetic difference. Take the following four factors in the causal sequence: (1) two X chromosomes → (2) being female → (3) being discriminated against → (4) being paid less. In this scenario, the causal link between (2) and (3) stands out for us so saliently as the *nervus explanandi* of the observed income difference that any account mentioning only factor (1) would strike us as a joke, stupidity, or perhaps just a non-standard causal attribution. And this is exactly the point that Jencks, Block, Sober, and others want to make: if the logic of heritability is so permissive that it counts even consequences of sexual or racial discrimination as "genetic" effects, this means that heritable differences (in the technical and counter-intuitive sense of behavior genetics) are not necessarily genetic effects *in our usual sense of the word*. Conclusion: heritability estimates, whatever their magnitude, are never by themselves evidence that genes play an important explanatory role simply because even variation that is 100 percent heritable (i.e., ultimately due to genetic differences) may well be 100 percent proximately explained by much more relevant environmental differences.

I will try to show that far from being so semantically perverse, the term "heritability," when properly understood, actually accords quite well with our commonsense etiological ascriptions. Before embarking on this task, however, I have to do some preliminary definitional work, and explain three different kinds of genotype–environment correlation (or covariance):[1] passive, reactive, and active. These terms were introduced in Plomin et al. 1977, although a very similar distinction was informally made earlier and often discussed through examples (see especially Jensen 1969b: 38; cf. Cattell 1963: 200; Cattell 1965: 110).

3.1.2 Three types of G–E correlation

Genotype–environment correlation is present when organisms with a given genotype tend to find themselves more often in one type of environment than do organisms with another genotype. I will follow the usual

---

[1] The correlation is standardized covariance.

practice and illustrate the three kinds of G–E covariance with hypothetical and non-tendentious examples involving IQ. First, if parents with higher IQ give to their children *both* genes for higher IQ *and* an intellectually more stimulating environment at home, this is passive G–E correlation. It is called "passive" because neither the children's behavior nor their geno-type is a causal factor that could account for the correlation. Second, if other people react to children with genotypically higher IQ by, say, impos-ing on them more intellectually demanding conversations and otherwise challenging and stimulating their ability even further, this is reactive (or evocative) G–E correlation. Finally, active G–E correlation occurs when brighter children themselves seek and eventually select those experiences and environments that they find specially stimulating.

Passive G–E correlation is not relevant for Jencks–Block–Sober argu-ment, as they all acknowledge (Jencks 1980: 725; Block 1995: 118; Sober 2001: 73). In this case, the common cause of the children's double (dis)advantage (genetic and environmental) is the parental genotype, which produces the environmental effect on children through the parental phenotype, and the genetic effect through meiosis and conception. So, the two (dis)advantages accrue to children in a way that is not mediated by their own genotype at all. One of the reasons why the critics of heritability exclude passive covariance from their consideration is that the presence of this kind of G–E covariance can be tested with traditional methods of behavior genetics (the adoption design), and there is consequently little temptation to treat this recognizably separate source of variance ana-lytically as a component of heritability. (For a limitation of this test see Rutter & Silberg 2002: 472.)

What about active and reactive G–E covariance? Well, behavior geneticists are indeed inclined to subsume some instances of the former under genetic variance. However, they are reluctant to do the same with the latter, particularly when it comes to those types of reactive covari-ance that would make the notion "heritability" misbehave conceptually in a way described by Jencks, Block, and Sober. (As explained above, the red-haired example is not a case of G–E correlation in the strict ANOVA sense, because in this case there is no separate, direct contribution to phenotypic variance from the genetic side, but I will ignore this subtlety here.)

In the case of active G–E correlation, the environments that lead to a phenotypic difference are *selected* by subjects themselves, whereas in reactive correlation they are *imposed* by others. Under some circum-stances this distinction may affect the way we decide to apportion causal responsibilities. Namely, the influences of those environments that are chosen on the basis of genotype are typically difficult to keep apart from the influence of genotype itself. In many instances the selection of these environmental influences can be plausibly regarded as just a way a genotype is expressed, and hence as "a more or less inevitable result of genotype" (Jinks & Fulker 1970: 323) or "merely an extension of the phenotype" (Eaves et al. 1977: 19). Therefore, phenotypic effects of such environments are indeed sometimes classified as heritable, on the grounds that they are practically inseparable from direct genetic effects and that they merely represent the self-realization of the genotype (Jensen 1969b: 39; Jinks & Fulker 1970: 323; Jensen 1973a: 54, 368; Jensen 1976a: 92–93; Rowe 1994: 90–92). In contrast, the key illustrations used by Jencks and others (red-haired children, sexual and racial discrimination) all involve reactive covariance in which environments are arbitrarily imposed from the outside. Now, the assertion that behavior genetics incorporates environmental influences into genetic variance in this sort of situation (imposed environments) surely cannot be justified by merely giving examples where behavior genetics does that in a different type of situation (in cases of active G–E correlation, i.e., selected environments).

I have to forestall a possible misinterpretation here. I do not want to suggest that active and reactive covariance differ intrinsically from one another with respect to their causal status, whereby, as it were, active G–E covariance should as a matter of principle be always treated as a part of genetic variance, whereas reactive G–E covariance ought never be subsumed under it. Rather, my claim is that G–E covariance is in the first place a source of variance that is *sui generis* and, as such, distinct from either the genetic or the environmental component of phenotypic variance (because G–E covariance involves *both* genetic and environmental influences). However, in certain specific situations researchers may wish to include the G–E covariance in genetic variance, but when they do that, they are typically guided by commonsense notions about causal attributions, rather than going against them. Namely, in some cases of active G–E covariance, G leads to E, which in turn leads to P, and all this unfolds in such a way that the genotype–environment correlation strikes us as just a self-actualization (or natural manifestation) of the genotype. Consequently, some behavior geneticists do tend to interpret phenotypic differences arising in this manner as resulting from genetic differences (that is, as being heritable) simply because they think that in that type of situation the role of the environment degenerates into its being a mere reflection of the genotype, or the way the genotype expresses itself: "To what extent could we ever get a dull person to select for himself an intellectually stimulating environment to the same extent as a bright person might?" (Jinks & Fulker 1970: 323).

The red-haired children example is totally different. In the cases of such blatant discrimination and abuse it would obviously be impossible to use the "self-realization" argument to incorporate the ensuing phenotypic variation into genetic variance. Actually, I am unaware that any serious scholar ever defended the idea that the indirect effects of Jencks-type scenarios should be treated as heritable.[2] On the contrary, a prominent behavior geneticist, a co-author of the first behavior genetics textbook, resolutely rejected the proposal, addressing this very issue head-on:

> In our human societies discriminatory practices are often based upon superficial physical characteristics or upon cultural stereotypes. In these instances a G–E correlation will result if, and only if, the criterion for discrimination is heritable in a genetic sense. If the criterion is a superficial physical trait (skin pigmentation, for example) it is of trivial behavioral interest. *Any correlation between it and behavior is logically attributable to environmental influences.* (Fuller 1979: 472 – italics added)

In the face of this explicit repudiation of the Jencksian construal of heritability by the leading authority in the field, the reader will surely wonder whether the critics actually offered any textual support for their "paradoxical" reading of that crucial concept of behavior genetics. As a matter of fact, there has been surprisingly little effort to document the charge with quotations from relevant sources. To make things worse, even in those few rare cases where the attempt has been made to provide evidence, a closer look into these "probative" texts always reveals that they were taken out of context and seriously misinterpreted. Here are some of these exegetical miscarriages.

---

[2] True, Loehlin et al. (1975: 87–88), discussing a fictional example of discrimination on the basis of eye color, say that since there the phenotype can be predicted from a knowledge of genotype, "in that sense it is not entirely incorrect to speak of a high heritability of this trait." However, the phrase "not entirely incorrect" clearly indicates that they do regard this way of speaking as misleading (it *is* incorrect but not entirely), and that they are uncomfortable with describing a phenotypic difference arising in this way as "heritable" or "genetic." Indeed, they consistently put scare quotes around these expressions in that context. (A similar case is treated in the same way in Lilienfeld & Waldman 2000.)

3.1.3 R. C. Roberts: the witness for the prosecution?

Both Jencks and Block make much of the following quotation from R. C. Roberts:

> it matters not one whit whether the effects of the genes are mediated through the external environment or directly, through, say, the ribosomes. (Roberts 1967: 218)

On the face of it, the quoted sentence does seem to express the view that it makes absolutely no difference for behavior-genetic analysis how genes cause phenotypic differences, the only important thing being that they are the ultimate causes. While Jencks quotes just that statement in isolation, Block gives a fuller version, and ironically the context he thereby supplies clearly reveals that even Roberts did not think that anything ultimately caused by genes should be automatically treated as a genetic effect. Here is the fuller version:

> The genotype may influence the phenotype either by means of biochemical or other processes, labelled for convenience as "development," or by means of influencing the animal's *choice of environment*. But this second pathway, just as much as the first, is a genetic one; formally it matters not one whit whether the effects of the genes are mediated through the external environment or directly, through, say, the ribosomes (Roberts 1967: 218 – italics added).

Roberts considers here just two kinds of genetic influence on phenotype, the one internal to the organism and the other via the animal's *choice of environment*. Apparently, the latter is *not* meant to include *any* indirect causation that starts with genotype and also involves environment, but only processes that involve genotype-*selected* environments (i.e., active G–E correlation). In other words, a more careful reading of Roberts's statement shows that it is restricted in content and that it does not apply to the cases of *imposed* environments at all. Therefore, it cannot be used (as it is by Jencks and Block) to impute to behavior geneticists the counterintuitive interpretation of heritability, according to which even effects of blatant discrimination would be heritable.[3]

---

[3] Block's attempt to fortify that imputation with a quotation from Jensen (Jensen 1973a: 54) fails for the same reason. Again, it escapes Block's attention that in the very passage he quotes, Jensen only says that what should be included in the genetic variance is that part of the G–E correlation that is "a result of the genotype's *selective utilization of*

In Roberts's article there is another passage (not quoted by Jencks and Block) that might appear to corroborate their diagnosis:

> the environment is *defined* as that which affects the phenotype indepen-
> dently of the genotype. If an effect stems from the genes, it is genetic; any
> other effect is an environmental one. (Roberts 1967: 218)

Again, a closer look changes the picture. First, in this context Roberts again seems to have in mind *active* G–E correlation because he speaks about "habitat *selection*" and "the animal's *choice* of environment."[4] Second, the whole paper is manifestly focused on research on animals where, for obvious reasons, the complex interactions characteristic for Jencks–Block scenarios (like the red-haired children example) do not come into play. Third, Roberts's position is subtler than could be judged from that single quotation. A few pages later he issues an explicit warning that the possibility of unrecognized G–E covariance could lead to overestimation of heritability: "The overriding concern at this stage is to avoid environmental sources of covariance that would lead to the wrong answer by inflating the estimate of the heritability . . . *There is no substitute for common sense in avoiding the pitfalls in this respect*" (Roberts 1967: 234–235 – italics added). And fourth, Roberts is an ill-chosen example for representing methods of behavior genetics simply because, strictly speaking, he is not a behavior geneticist at all, but a quantitative geneticist who, as he himself says, was just asked on that occasion to "look over the wall into the field of behavior genetics."

3.1.4 The concept of "environment": broad and narrow

Jencks's criticism of behavior geneticists is basically that they have changed the ordinary meaning of "environment" (Jencks 1980: 726). Allegedly, they narrowed the connotation of the term so that it came to include only the environments that are *not* correlated with genotype. Jencks also claims that, as the consequent terminological ambiguity is not

---

the environment" (italics added). Clearly, Jensen had in mind the active, not reactive correlation. (Cf. Jensen 1976a: 92–93.)

[4] Later he explicitly argues that passive G–E correlation should be excluded from genetic variance (Roberts 1967: 234–235). He never discusses anything that we would now call "reactive covariance," which actually represents the key issue for Jencks and Block.

readily recognizable, heritability information will frequently appear more interesting than it actually is:

> Narrowing the definition (while retaining the term itself) is certain to mislead all but the most attentive and sophisticated readers. Indeed, it is only a slight exaggeration to say that narrowing the definition of a term that has traditionally had a very broad meaning is *meant* to mislead – that is, meant to make one's results sound more significant than they really are. (Jencks 1980: 726)

One would expect such a sweeping and damning criticism to be documented with a long list of references illustrating this massive semantic shift in the nature–nurture debate. However, Jencks gives only one example where the redefinition of "environment" is openly defended. Rather surprisingly, the source in question is actually his own article from 1977 (co-authored with Marsha Brown). Needless to say, the fact that in that article Jencks indeed introduced the idiosyncratically narrow definition of "environment" doesn't begin to show that the same conception has been widely shared by others or that it is endemic to the whole field. Moreover, the suggestion from that article that "environment" should be treated as a remainder term after genetic effects are directly estimated was explicitly rejected in a seminal paper on genotype–environment correlation and in a classic book on racial differences in intelligence:

> If some appreciable fraction of the variation of a trait is due to the covariation of genes and environments, *that portion of the variance can be assigned neither to heredity nor to environment* – it is attributable jointly to both. If one's analytic method proceeds by estimating the genetic effect directly and the environmental effect by subtraction, the presence of GE correlation could lead to a substantial underestimation of the latter. (Loehlin & DeFries 1987: 264 – italics added)

> The added component of variance [due to G–E association] cannot logically be assigned to either genes or the environment: it is a result of the association of their separate effects. (Loehlin et al. 1975: 78)

Although Block (1995) cites the paper by Loehlin and DeFries, he seems to be unaware of their unequivocal refusal to subsume G–E correlation under genetic variance, which directly contradicts his allegation that in behavior genetics, "if there is a genetic difference in the causal chains that lead to different characteristics, the difference counts as genetically caused even if the environmental differences are just as important."

That the methods of behavior genetics are not as crude as painted by Block is further confirmed in an article from a book that also happens to be in his list of references: "The issues of genotype–environment correlation and interaction are complex – *not nearly as simple as saying that their effects are incorporated wholly in estimates of genetic variance*" (Plomin 1987: 43 – italics added).

3.1.5 Does twin research necessarily underestimate environmental influences?

Here is how Block and Dworkin try to justify the claim that methods of behavior genetics necessarily generate counter-intuitive causal attributions in that they would count as heritable even those effects that we would usually classify as environmentally caused:

> To see how methods of estimating heritability automatically attribute to genes the effect of variance jointly caused by genes and environment, consider how this phenomenon [an imagined example of all red-haired children being given a near-starvation diet] would affect a study of one-egg twins reared apart. A pair of red-haired twins *both* will be ill-nourished in their respective adoptive homes, while a pair of blonde twins will both be adequately nourished. Thus the pervasive discrimination will contribute to the correlation between the twins' heights, and thus contribute to the assessed heritability of height in the population. (Block & Dworkin 1976b: 481)

This is a serious distortion of the way twin studies are used to estimate heritability. Yes, the similarity of monozygotic twins reared apart (MZA) is indeed taken to be a direct measure of heritability, *but only on the assumption that relevant environments of these twins are uncorrelated.* If this assumption is not satisfied or if there is some suspicion that it is not, no one would be so foolish as to continue using MZA phenotypic similarity to infer heritability. On the contrary, as has routinely been emphasized in the literature, the inference is considered legitimate *only if* there are no common environmental influences that could explain the concordance between the MZA twins. The same applies to the method that compares MZ and DZ (dizygotic) twins. Already in the first book on behavior genetics ("the field-defining book," according to Plomin 1987: 110), it is clearly explained that Holzinger's formula[5] for calculating heritability

---

[5] $h^2 = (r_{mz} - r_{dz})/(1 - r_{dz})$. Here, r_{mz} and r_{dz} are phenotypic correlations between monozygotic and dizygotic twins, respectively.

depends on the assumption that "MZ and DZ pairs are treated nearly enough alike so that environmental differences between cotwins are equal for both types" (Fuller & Thompson 1960: 113; see also McClearn 1963: 196; DeFries 1967: 328–329; Jensen 1969b: 51–52; Fulker 1974: 94–95). This requirement has actually become an indispensable part of the twin research design and is known as "the equal environments assumption" (e.g., Plomin et al. 1976; Rowe 1983; Plomin 1987: 43; Rowe 1993; McGuffin & Martin 1999; Plomin et al. 2001: 80–82).

All this shows that phenotypic similarity of MZA twins is *by no means* "automatically" counted as a measure of heritability. Also, there is nothing that should put the correctly interpreted methods of estimating heritability "in violent conflict with our ordinary socially important ideas of causation" (Block 1995: 116). For, if MZA twins have no more similar environments than pairs of randomly chosen individuals, this is indeed usually treated as a good reason to think that the degree of observed similarity between them correctly reflects heritability. Why would common sense be offended by that? If, on the other hand, MZA twins happen to have more similar environments than others, common sense would advise against deriving any conclusion about heritability because then the twins' phenotypic similarity might be the result of that greater similarity of environmental influences. And behavior geneticists would wholeheartedly agree.

As an analogy, consider the following "method" for recognizing murderers: if someone is convicted of murder in a court of law, conclude that the person is a murderer. This is a perfectly reasonable rule of inference. It would be ridiculous to criticize it as being "violently in conflict with common sense" on the grounds that, allegedly, the application of this rule would force us to regard someone as a murderer even if, for example, the trial judge was bribed to convict him. Of course, no such consequence follows. Our use of that rule in any particular case is based on the assumption that the procedure was fair, and that, among other things, the judge was not bribed. Clearly, this assumption can always turn out to be false, but the point is precisely that if we *know* (or suspect) that it is false, we would never apply that rule.

It is the same with heritability. Take a method for estimating heritability, very roughly described as follows: if there is a strong correlation between MZA twins with respect to a given trait, conclude that the heritability of that trait is high. This method obviously makes sense only as long as we have reason to believe that there are no strong confounding influences of non-genetic factors (like the increased similarity of MZA twins' trait-relevant environments). As soon as we start to have doubts about this, the inference breaks down and we realize that knowledge about heritability cannot be obtained with that method. Briefly, if the crucial assumption is compromised and there is significant G–E correlation (as in the case of red-haired twins) the result is not, as Jencks and Block argue, that the method delivers a heritability estimate with a counter-intuitive causal attribution. No, the result is that if the assumption is false, this particular method just becomes inapplicable: the MZA's phenotypic correlation simply ceases to be a straightforward measure of heritability.

But what about the assumption itself? Can we ever make reasonably sure that G–E correlation is absent? Jencks and Block (and many others as well) argue that we cannot. Note that this is a completely different argument from the one discussed until now. It addresses empirical matters, and leaves semantic and definitional issues behind. This argument will be discussed in section 3.2.

3.1.6 What Falconer did not say

Elliott Sober also thinks that the semantics of heritability conflicts with common sense:

> Quantitative geneticists differ from the rest of us in the way they tend to use the term "environment," and this difference in usage will probably persist. This means that when quantitative geneticists say that the variation in some phenotype has a genetic component, the rest of us must be very careful. (Sober 2001: 75; cf. Sober 2000: 366)

In discussing the example with the abuse of redheads, he warns that geneticists are forced to describe that kind of situation in a bizarre way, and that they have to interpret the lowering of IQ, counter-intuitively, as a genetic effect: "Quantitative geneticists do not regard abuse as an environmental factor ... The lower IQ of redheads ... is said to be genetic, rather than environmental, on the grounds that individuals experience abuse because of their genes" (Sober 2001: 73). He refers at this point to Falconer's *Introduction to Quantitative Genetics*, directly in the context of the redheads example, which might be read as suggesting that Falconer both discussed that very example and claimed that in such cases the result of the abuse should be regarded as a genetic effect. But in fact Falconer did not address that kind of issue at all. His treatment of G–E correlation is very general and superficial (and is limited to just one paragraph).

The only two examples of G–E correlation that he mentions deal with cows' milk yield (where cows are fed according to their yield, the better phenotypes being given more food) and with human intelligence (where genotypically mediated phenotypes of the parents affect the environment in which the children grow up). In the latter case, Falconer explicitly refuses to count an individual's environment as a part of its genotype, "because the environmental effects on the children are not a consequence of their own genotype, but of their parents' genotype" (that is, because it is *passive* G–E correlation). In the former case (with cows), which in our taxonomy would correspond to reactive G–E correlation, he does include the environmental effect in genetic variance, but *only* under the assumption that the G–E correlation is "in practice unknown." So, there is no suggestion, even here, that the effects of those environments *known* to be ultimately but just superficially associated with a given genotype should be imperatively attributed to that genotype (as Jencks, Block, and Sober would have it).

Most importantly, although in the context of a selective breeding of cattle it might perhaps be all right to regard all consequences that ultimately originate from genetic differences as genetic effects (because the purpose here is to make the genotypic superiority manifest itself as much as possible), it by no means follows that the same causal analysis would be acceptable in incomparably more complex situations involving people's behavior and intricate social interactions. Certainly, nothing Falconer says warrants the belief that he would endorse such an extrapolation. Besides, Falconer seems to be the wrong source for this discussion. To know how researchers would describe the redheads example and similar cases, it doesn't seem appropriate to refer to a standard introduction to quantitative genetics, which will necessarily be too general and too coarse-grained to be a useful guide for situations of that level of social complexity. Rather, one is much better advised to look into those (numerous) publications in behavior genetics that try to deal precisely with such tangled mixtures of genetic and social influences. My search of this literature has uncovered no evidence that would corroborate the "paradoxical" reading of heritability advocated by many critics of behavior genetics.

In fact, just the opposite is true. In a paper that addresses the very issue of estimating heritability in the presence of G–E correlation, it is claimed that under these conditions "the concept of heritability may not be informative or relevant" (Emigh 1977: 514), that it becomes "uncertain how to interpret the partition[ing] of variance" (508), and that what is a reasonable measure of heritability in those cases where G–E correlation is

102

small or non-existent *"must not be applied blindly to all situations"* (509 – italics added).

So, the flaw that the critics attributed to the concept of heritability was of their own making. They first applied the notion to the situations outside of its proper range (against the explicit warning in the literature not to do this), and then they concluded that the concept was somehow defective or intrinsically confusing.

3.1.7 No "violent conflict" with common sense

What then about this disciplinary revision of the concepts "environment" and "heritable" hypothesized by Jencks, Block, Sober,[6] and others? Have these words really undergone a momentous change of meaning in behavior genetics, whereby "environment" so shrank in content that it started to exclude many bona fide environmental effects, while "heritability" correspondingly expanded to the point of getting, as Block would put it, violently in conflict with our ordinary socially important ideas of causation (Block 1995: 116)? I think that this whole story is a myth. It is, for example, clearly belied by the terminological practice in that large and notorious segment of the heredity–environment controversy, the race and IQ issue. In this debate, literally no one would say that explaining the black–white IQ gap by appeal to discrimination would show that the difference is heritable, although this is exactly how the situation should be described by using the "narrow definition" of environment. On that definition, if discrimination is correlated with a genetically mediated difference of skin color, its effects should not be regarded as an environmental influence (analogously to the red-haired children example). But no one is tempted to use the narrow definition here, and for a very good reason. Labeling the IQ gap resulting from discrimination as heritable would defeat the purpose of the term. It would make the IQ difference between the races heritable by definition, and then one would have to invent a new concept (*"really* heritable"?) to discuss the central question whether genes are explanatorily involved in a stronger and more interesting way than by just giving rise to a superficial characteristic which in turn becomes a target of discriminatory practices.

Besides, contrary to what Jencks and others assert, there is no reason why the notion "heritability" should be extended to cover such cases.

---

[6] Jencks speaks about "redefinition," and Block and Sober about "tacit convention" and "convention."

True, there is a tendency to take some part of the variance due to G–E correlation and divide it between genes and environments. Thus active G–E covariance is occasionally subsumed under genetic variance, and passive covariance is assigned to the environmental side of the equation. However, it is important to stress that this redistribution is not a necessary consequence of some esoteric methodology for calculating heritability. Rather, it is a practical decision primarily guided by an attempt to follow the commonsense way of apportioning causal responsibility. If there happens to be any doubt about how to classify G–E correlation, the regular fallback position is just to treat it as a distinct component of variance, separate from heritability and environmentality. But even in those cases where a researcher opts for subsuming it under one of the two main effects, this is always just a pragmatic decision. There is no "correct" answer that is dictated, as it were, by the logic of heritability (cf. Jensen 1969b: 39; Jensen 1981a: 121).

To see that the way of causal attribution in behavior genetics is not really anomalous or aberrant, look at one of the many examples where common sense would treat indirect causation in basically the same manner. Suppose that in a certain region a positive correlation is observed between rainy weather and the number of injured drivers. Consider the following two possible explanations of that correlation: (1) rain makes roads slippery, and this leads to injuries; and (2) there is a crazy gang of psychopaths who go out during rain to drop bricks on cars from highway bridges, and this leads to injuries.

In (1) – which corresponds to the effects of self-selected environments – it seems perfectly natural to say that rain causes injuries, essentially because rain is regarded as almost inseparable from roads becoming slippery, which in turn is an immediate cause of injuries (rain → slippery roads → injuries). However, in (2) – which corresponds to the effects of imposed environments – it sounds wrong to say that rain causes injuries, although here, again, rain *is* an indirect cause of the greater number of injuries (rain → psychopaths with bricks → injuries). Why? I submit that one reason is that now rain has only a tenuous link with the mediating cause, and consequently the mediating cause gains in independence, salience, and explanatory importance. As a result, rain can no longer account for the effect on its own, i.e., without mentioning the middle link. Put differently, we would not regard the bricks falling from the bridges as a "natural" manifestation of rain.

Another contrast to (1), and an additional similarity to the "imposed environments" scenario, is that in (2), the issue of blame is involved, and in such cases we usually want to focus on the human action responsible for the final outcome, and not just on the antecedent conditions of that action. This is all admittedly pretty vague, and I am not sure how intuitions underlying our different approach to these two kinds of cases should be refined further and made more precise. Fortunately this doesn't really matter, for I only want to claim that in dealing with G–E correlations, behavior geneticists are by and large guided by the commonsense considerations about causality, with all their characteristic vagueness and ambiguities.

It is somewhat ironic that the phenomenon of G–E correlation has been used to argue that the notion of heritability is, as Block says, in violent conflict with common sense. For, in fact, behavior geneticists rely on nothing but common sense whenever they subsume (active) G–E correlation under heritability, i.e., whenever they decide to abandon the default option of treating G–E correlation as a separate, *sui generis* source of phenotypic variance (which is neither genetic nor environmental).

3.2 THE METHODOLOGICAL ROUTE: TRACING THE PATHS OF CAUSALITY

Running in parallel with the conceptual criticism of heritability (and not clearly separated from it) there is another strand of argument in the literature that is more empirically oriented. Rather than criticizing heritability for "redefining" the common notion of environment or "distorting the ways in which we normally think about causation," sometimes the main complaint seems to be that heritability lumps together genetic and environmental influences merely because in practice there is no feasible way to keep these two kinds of effects apart.

Since there is no practical method for separating the physical and social effects of genes, heritability estimates include both. This means that heritability estimates set a lower bound on the explanatory power of the environment, not an upper bound. If genetic variation explains 60 percent of the variation in IQ scores, environment variation *must* explain the remaining 40 percent, but it *may* explain as much as 100 percent. (Jencks 1992: 107)

Where does the "gene–environment covariance" show up in heritability calculations? Answer: active and reactive effects that we don't know how to measure inevitably are included in the genetic component. (Block 1995: 120)

Let us compare the two lines of argument. In the conceptual version, the alleged inclusion of indirect genetic effects in heritability was represented as being based on the idea that anything caused by genetic differences should be regarded as a genetic effect. In the methodological version, however, the inclusion is not pictured as a principled decision but more as a desperate move resulting from an epistemological predicament, i.e., inability to differentiate empirically between direct genetic effects and indirect effects via environment. In the conceptual approach even *known* indirect effects are classified as heritable (which creates a conflict with common sense), whereas in the methodological approach this applies only to *unknown* indirect effects. Clearly, rebutting the conceptual argument is not enough to remove methodological doubts that the phenomenon of G–E correlation raises about heritability. For, even if heritability were defined in such a way as to exclude arbitrary indirect effects like the red-haired children situation, what would be the use of such an analytically immaculate concept if it could never be applied in practice?

The dilemma seems to be this. On one hand, if heritability is taken to include arbitrary indirect effects, then the concept can be empirically applied without difficulties, but its usefulness is very dubious because it represents a hopeless jumble of genetic and environmental effects (in the ordinary sense of these words). On the other hand, if heritability is taken to exclude arbitrary indirect effects, then the concept is semantically well-behaved but now it becomes doubtful whether we can ever get reliable information about so-defined heritability because, the argument goes, the indirect effects that we want to shut out are not detectable with the standard methods of behavior genetics.

Indeed, one of the most frequent methodological criticisms of heritability is to say that the indirect genetic effects produced by the correlation of genetic and environmental differences make it impossible to get a meaningful estimate of heritability:

> When this condition [the absence of genotype–environment correlation] is not satisfied, the contributions of interaction to phenotypic variances and covariances cannot, in general, be separated from the contributions of genotype and environment, and heritability analysis cannot, therefore, be applied meaningfully. (Layzer 1974: 1263)

> the very existence of genotype–environment correlation precludes the valid statistical estimation of the genotypic, environmental and interaction contributions to the phenotypic variance. That is because correlation makes

it *impossible* to know how much of the phenotypic similarity arises from similarity of the environment. (Feldman & Lewontin 1975: 1164 – italics added)

I found that the theoretical consequences of genotype–environment correlation are in fact extremely serious. When such correlations are present, as they undoubtedly are for most phenotypically plastic traits in natural human populations, they make it impossible to estimate heritability – or even to define it in a satisfactory way. (Layzer 1976: 209)[7]

These are all very strong claims. If accepted, they would amount to a kind of "impossibility proof" for measuring heritability of human psychological differences. (As mentioned before, it is usually conceded that in animal research the problem of G–E correlation can be overcome by breeding experiments or by randomizing environments.)

3.2.1 The equal environments assumption

Discussing a possible involvement of genes in explaining the difference in human sexual orientation, Philip Kitcher states that inferring heritability from the greater phenotypic concordance of monozygotic twins is dependent on the premise that their environments are not more similar than those of an average pair of siblings. He then explains why he thinks that this premise "is almost certainly false":

As behavioral geneticists know all too well, monozygotic twins share environments that are far more similar than the environments of ordinary siblings or of fraternal twins. The similarity of the environments confounds the effects of genetic identity. The grand (and unsurprising) conclusion is thus that individuals who are more alike genetically and who are reared in more similar environments are likely to behave more similarly than those who are genetically more diverse and who are reared in more varied environments. (Kitcher 1985: 248)

There are four problems with Kitcher's reasoning. First, the crucial premise on which the twin research hinges is incorrectly stated. Contrary to what he says, if the environments of MZ twins are overall more similar

---

[7] Layzer's criticism of heritability has been widely accepted in philosophy of science, starting with the inclusion of one of his papers in the Block–Dworkin anthology, and continuing to have a lot of influence afterwards. For instance, Arthur Fine states that "one of the most thoughtful contributions to the public debate on [the nature–nurture issue] was made by David Layzer, who identified the conditions required for an adequate study" (Fine 1990: 95).

than those of DZ twins, this by itself is *not* enough to undermine the inference about heritability. What matters, and matters only, is whether this increased environmental similarity of MZ twins also *causes* the twins to be increasingly similar phenotypically. If there is a way to show that the greater environmental similarity of MZ twins has no impact on the phenotype, then its presence would be no obstacle to obtain the heritability estimate using the twin method. So, even if Kitcher is right that (1) MZ twins have environments that are more similar than those of an average pair of siblings, it does not follow that (2) MZ twins have *phenotype-influencing* trait-relevant environments that are more similar than those of an average pair of siblings. But (2) is actually the premise in the heritability inference, not (1). In other words, if (1) is true and (2) is false, the calculation of heritability on the basis of twin data can proceed smoothly.

Second, MZ twins do *not* always have more similar environments than DZ twins. It is particularly with respect to prenatal environments that MZ twins are sometimes *more different* than DZ twins. For example, a substantial proportion of MZ twin-pairs has the third blood circulation system (the so-called "Chronic Fetofetal Transfusion Syndrome") that leads to serious developmental asymmetries and birth-weight differences, which all tend to increase the MZ environmental variance (cf. Hay 1985: 227). Relying on this kind of empirical evidence, some authors have claimed that the twin research method actually *underestimates* the true magnitude of heritability, rather than overestimating it (see Price 1950, Munsinger 1977, and references in Bouchard 1998: 270). I do not want to endorse that view here. My point is only that the situation is much more complicated than Kitcher imagines.

Third, Kitcher's quick repudiation of the twin method creates a completely false impression about behavior genetics as a field where methodologically incompetent people continue with obviously fallacious inferences, despite being constantly warned about it. The reader gets no information that, in reality, there has been a massive research effort to find out whether "the similarity of the environments confounds the effects of genetic identity." The empirical results go flatly against what Kitcher so dogmatically asserts. Rather than being "almost certainly false," the assumption that trait-relevant environments are *not* more similar in twins than in other siblings (the so-called "equal environments assumption," or EEA) turns out to be strongly confirmed in empirical studies about many traits. Typically, even when MZ environments are more similar, it just happens that this increased environmental similarity is not translated into increased phenotypic similarity. (Various methods of testing EEA will be briefly described later.)

And fourth, many environmental factors cannot be invoked to explain the phenotypic correlation of MZA (monozygotic twins reared apart) simply because for many traits the impact of these factors on phenotype has been shown to be weak or non-existent in studies of biologically unrelated children who were adopted and raised in the same home. In other words, since the correlation between family environment and phenotype in adulthood has been found to be close to zero, this puts a severe constraint on environmental explanations of MZA similarities.[8] This is one of the reasons why Kitcher is seriously wrong when he seems to suggest that the very high phenotypic concordance of monozygotic twins could easily be explained by their environmental similarity.

In criticizing behavior genetics Kitcher gives no reference to the works of behavior geneticists, but there is a reference to the book *Not in Our Genes*, where the same objection was raised (Rose et al. 1984: 116). It is understandable that behavior geneticists become frustrated when they see this type of objection spreading around and their research being quickly dismissed on that basis, without the critics citing (let alone discussing) any of the numerous empirical studies in this area. David Rowe commented in desperation: "For many phenotypes, leveling this criticism at twins studies has been a red herring – of little real merit, but unfortunately effective, because by repetition many people have come to believe it (without much serious reflection on their part)" (Rowe 2002: 21).

A non sequitur similar to Kitcher's is committed in one of the most widely used textbooks of genetics: "Identical (monozygotic) twins are generally treated more similarly to each other than fraternal (dizygotic) twins. People often give their identical twins names that are similar, dress them alike, treat them identically, and, in general, accent their similarities. *As a result*, heritability is overestimated" (Griffiths et al. 2000: 756 – italics added). Again, from the similar treatment alone the overestimation of heritability does not necessarily follow "as a result." For this result to follow, the similar treatment would also have to have an effect on

---

[8] Since this research design (a phenotypic comparison of biologically unrelated individuals who were raised together) is such a powerful method of testing the trait-relevance of many environmental influences, Bouchard thinks that it is a mystery that this type of study has not been used more often. He said that it is "almost as though psychologists did not wish to collect data using a sample that would refute their favorite hypotheses" (Bouchard 1997: 136).

phenotype, and of course, whether this will happen is always an open empirical question.

Elliott Sober also moves too fast in this context. He says (Sober 2001: 68) that inferring heritability from twin data depends on the equal environments assumption, which he represents by the following proposition about two variances (for monozygotic and dizygotic twins): $V_e(MZ) = V_e(DZ)$. He then suggests that the proposition may be false if, say, "parents treat identical twins more similarly than they treat fraternal twins" (Sober 2001: 69–70). But again, differential treatment is by no means sufficient to refute the equal environments assumption. Sober seems to have overlooked here that in the above proposition, the term "V_e" (the way he defined it himself) does not refer to within-twin environmental variance as such but to the environmental part of within-twin *phenotypic* variance. Therefore, the assumption is not so easily discredited since it is not about the mere existence of environmental dissimilarity but about its *causal influence on the trait variation*.

Sober makes the same inferential leap when, from the assumption that twins are reared in similar environments (through being adopted either into the homes of relatives, or by people with high socioeconomic status or SES), he concludes that within-twin V_e will be smaller than V_e in the general population. However, this does not follow at all, because "V_e" is the environmental part of *phenotypic* variance, and it is far from clear that home environment and SES have a systematic influence on psychological traits. In the case of IQ in adulthood, there is actually strong empirical evidence that they do not.

At another place, again, Sober uses the premise that twins' environments are somewhat similar to conclude that some of the phenotypic similarity between the twins will be non-genetic (Sober 2000: 364). But, of course, the conclusion follows only on the additional assumption that the similar environments are trait-relevant.

Among many others, Garland E. Allen also makes the incorrect claim that any heritability claim is meaningless unless we know that the environment for the whole population is uniform throughout (Allen 1994: 183).

3.2.2 *Ignorance of causal mechanisms*

Here is how Ned Block explains why direct genetic influences and indirect genetic (environmentally mediated) effects cannot be disentangled from one another in the case of IQ differences:

so little is known about the genetic mechanisms underlying the development of IQ that no estimate of the importance of indirect genetic effects
can be made. . . . no one knows *how* to separate the variance due to indirect
genetic effects from the variance due to direct genetic effects, at least within
the constraints on human experimentation. Such a separation would involve
investigation of the details of the mechanisms by which genes affect psychological characteristics, a task which is well beyond present knowledge.
(Block & Dworkin 1976b: 482)

In the case of IQ, no one has any idea how to separate out direct from
indirect genetic effects because no one has much of an idea how genes and
environments affect IQ. (Block 1995: 117)

But reactive and active covariance cannot be measured without specific
hypotheses about how the environment affects IQ. And it is just *a fact
about IQ that little is known* about how the environment affects it. So reactive and active covariance is on the whole beyond the reach of the empirical methods of our era's "behavior genetics," for those methods do not
include an understanding of what IQ *is* (e.g., whether it is information-
processing capacity) or how the environment affects it. (Block 1995:
119)

According to Block, in order to uncouple direct genetic effects on IQ
from influences of G–E correlation it is imperative to know *how* genes
and environments affect IQ. If this is really a precondition for disentangling genetic and environmental contributions to phenotypic variance, the
prospects for behavior genetics are indeed bleak. For, the whole point of
heritability analysis is that it promises to partition the variance between
nature and nurture *despite total ignorance of specific mechanisms responsible for phenotypic differences*.

Therefore, if it is true that nature cannot be separated from nurture
until the causal details of developmental processes are already in full view,
the enterprise will lose much of its interest because at the time when that
late stage is reached the information about heritability would be largely
overshadowed by the more fine-grained knowledge about causal mechanisms. Analysis of variance is essentially a *faute de mieux* methodology
that gives the most valuable results precisely in those situations where
more in-depth approaches falter because the minutiae of causal process
are still inscrutable.

The main problem with Block's criticism is that he is trying to prove
too much. When he says that "in the case of IQ, no one has any idea how

to separate out direct from indirect genetic effects because no one has much of an idea how genes and environments affect IQ" (Block 1995: 117), or that "without an understanding of how the environment affects IQ, we simply have no way of determining how much of the variance in IQ is indirect genetic variance" (Block 1995: 119), or that "so little is known about the genetic mechanisms underlying the development of IQ that no estimate of the importance of indirect genetic effects can be made" (Block & Dworkin 1976b: 482), this reasoning is quite *general*, and it doesn't seem to be restricted to the IQ debate, nor, for that matter, to behavior genetics.

Block definitely seems to be committed here to the view that whenever, in a non-experimental context, we do not know how two variables X and Y affect the third variable Z, we will be unable to make any progress in separating direct and indirect effects of these two independent variables on Z. This is an exceptionally bold and sweeping claim about causal analysis, and it invites two short comments.

First, if accepted, it would send big ripples through social science, and a lot of research that is usually regarded as sound would stand condemned.[9] Second, an article about heritability is surely not an appropriate place to launch that globally pessimistic message. To attack behavior genetics with such a methodological weapon of mass destruction is a fallacy known as *qui nimium probat nihil probat* ("whoever proves too much, proves nothing").

In a similar attack that hits the intended target only by causing collateral damage of colossal proportions, Clark Glymour argues that the most effective criticism of *The Bell Curve* would raise the bar for a legitimate causal influence so high that it would entail rejecting large parts of the social sciences and their methods (Glymour 1997; Glymour 1998). Glymour defends that criticism, and is ready to pay the price. But in that case the titles of the articles in which he develops this line of thought ("Reflections on *The Bell Curve*," "Reflections on Science by Observation and *The Bell Curve*") are quite misleading and inappropriate, because they falsely suggest that a particular target is being aimed at. In fact, Glymour's objections resemble cluster bombs in being indiscriminately destructive over a very wide area, and they therefore apply with equal

---

[9] "If knowledge of mechanisms were required prior to investigation of relationships between predictor and outcome, how much of behavioral science would be disallowed?" (Turkheimer et al. 1995: 149).

force to environmentalists like Jencks and Flynn as they do to hereditarians like Herrnstein and Murray.

3.2.3 Testing EEA

I want to show that even without knowing *how* the environment affects a given trait, the presence/absence of G–E covariance *is* amenable to empirical investigation. How then, for instance, can we know that the phenotypic similarity of monozygotic twins reared apart is not due to their being treated similarly by others? That is, how can we rule out the hypothesis that the greater trait concordance of MZA twins is produced by their sharing some heritable and recognizable but not directly trait-relevant characteristic (say, the similarity of physical appearance), which then makes other people treat them similarly (say, giving them more/less trait-enhancing stimulation), which in turn happens to be the real cause of their increased trait resemblance?

This kind of scenario involves four causal factors: G (genetic difference) → C (characteristic that is not directly trait-relevant) → E (environmental influence) → P (trait difference). Whether something like this really occurs can be tested in different ways.

(i) The scenario implies that C will be correlated with P: genetically related individuals are claimed to be more alike with respect to both C and P. One way, then, to sort out the causal connections is to check whether the correlation between C and P will still exist when G is controlled for. If the correlation disappears, this would strongly suggest that G is actually the common cause of C and P, and that G is *not* causing P via C. The four-link scenario would be undermined.

Essentially this is what Sandra Scarr did when she tested a very popular environmentalist hypothesis (H) in which the externally perceived identity of appearance of MZ twins plays the role of C. According to H, MZ twins are more similar with respect to some phenotypic trait(s) P merely because their genetic relatedness (G) makes them look similar (C), which in turn makes others treat them similarly (E), and this similar treatment by others (E) is the crucial causal influence on P. Scarr (1968) and Scarr and Carter-Saltzman (1982) compared MZ twins (who are 100 percent G-similar and strongly C-similar) with those DZ twins who, being almost indistinguishable from one another, are incorrectly classified as MZ twins (who are strongly C-similar despite being only 50 percent G-similar). Obviously, H predicts that these two kinds of twins should exhibit an increased degree of twin-to-twin similarity with respect to P, because they share the same degree of C (the similar external appearance), the hypothesized "main cause." But although many phenotypic traits were explored, the prediction was disconfirmed: despite their increased perceived similarity, DZ twins incorrectly classified as MZ were *far less* similar phenotypically than real MZ twins.

(ii) An alternative testing method would be to check whether another prediction of the above four-link scenario holds, namely whether P-differences *within* twin pairs are systematically related to their E-differences. For example, if it is twins' being dressed alike that makes them phenotypically similar, it would follow that those twins who are not dressed alike should show less trait similarity than those who are. John Loehlin and Robert Nichols (Loehlin & Nichols 1976) conducted such a study with a large sample, and the result was essentially negative. They collected information about twins' different treatments (as reported by their mothers), but the average correlation between a composite measure of these various possible E-factors and a number of psychological traits was very close to zero (0.056).

(iii) Yet another approach is based on a simple idea (Loehlin 1992: 109; Neale & Cardon 1992: 223): take a number of pairs of MZ twins reared apart, and then just check whether there is a correlation between a given environmental measure of one twin and the other twin's phenotype. If there is no correlation, this would indicate that the MZA phenotypic similarity is *not* mediated through this particular environmental measure. (In active and reactive G–E correlation the causal relation is G → E → P, and since the twins have the same genotype, G could not influence P via E if there were no cross-correlation between E and P.) Plomin (Plomin et al. 2001: 311–312) used a similar research design in the context of the Colorado Adoption Project but with parent–child pairs instead of twins. Only meager evidence was found for reactive or active G–E covariance: just a few and fairly low correlations were found between the biological mother's personality traits and the adopted children's environmental measures.[10]

---

[10] A number of personality traits were simultaneously considered, and their relation with a wide range of environmental characteristics included in the so-called HOME scale (Home Observation for Measurement of the Environment). HOME contains information about things like the emotional and verbal responsivity of the caregiver, avoidance of restriction and punishment, organization of the physical and temporal environment, provision of appropriate play materials, caregiver involvement with the child, and opportunities for variety in daily stimulation.

(iv) If it is the environments shared by MZ twins that cause their phenotypic similarity, one would expect that their similarity would decrease with earlier age of separation, lower degree of contact, and less time spent together. But the massive study of Swedish twins found no effects of this kind (Lykken 1995: 78).

(v) Finally, the most effective way to empirically investigate the issue of G–E covariance is the multivariate genetic analysis. Here, the basic strategy is to look simultaneously at the impact of genetic differences on two variables (environment and phenotype) in the attempt to estimate whether the two effects overlap (and if yes, to what extent). Explorations with this method have not unearthed strong G–E correlations of the kind envisaged by Block and others, and certainly nothing of the order that could put into serious doubt the accepted high heritability estimates for IQ and many other psychological traits.

For those who would like to get more detailed information about how EEA is empirically evaluated in practice, the literature is huge. Here is just a random selection of several recent studies: Kendler & Prescott 1993; Hettema et al. 1995; Kendler et al. 1999, 2000; Klump et al. 2000.

3.2.4 *False support from behavior genetics*

Of course, expressing confidence that current methods are sufficiently sensitive to detect sizable G–E covariance does not entail the unrealistic belief that every single manifestation of that phenomenon will be known in its entirety. Interestingly, when some researchers say exactly this, and warn in a low tone of voice that reality is too complex to expect an absolute and exhaustive knowledge of all the causes of phenotypic differences, Block misrepresents this as their agreement with his quite radical claim that reactive and active covariance are "beyond the reach of the empirical methods of our era's behavior genetics." Block first quotes from an important paper by Plomin, DeFries, and Loehlin:

> Because it is not possible to measure all aspects of the environment (including everybody and everything) that might correlate with children's genotypes, it will probably never be possible to assess completely the effects of active and reactive genotype-environment correlations. (Plomin et al. 1977: 321)

. . . and then he superimposes his interpretation of what they wanted to say:

The upshot is that there may be a *large* component of heritability due to indirect genetic effects, including (but not limited to) gene–environment correlation, that is outside the boundaries of what can be measured given the mainly atheoretical approach that is available today. (Block 1995: 120 – italics added)

But this is wrong. The "upshot" of Plomin et al. is most assuredly not that there may be a *large* component of heritability due to indirect genetic effects that is undetectable by today's methods. For their words quoted by Block are immediately followed by a statement that directly contradicts his interpretation:

However, if the effects of genotype–environment correlation are *important* and *pervasive*, the test that we have suggested should detect them. (Plomin et al. 1977: 321 – italics added)

So, their point is quite clearly that if the effects are large enough they will be detectable. Block thinks, on the contrary, that the empirical uncertainty about the size of G–E correlation is without bounds, and that its effect "could, for all we know, largely account for the heritability of IQ" (Block 1995: 119).

This suggestion that the usually high estimate of heritability of IQ could be explained away in its totality as being due to the indirect genetic effects (of the G–E covariance type) has always been deemed by behavior geneticists extremely implausible. For instance, David Fulker, one of the most methodologically sophisticated scientists in the field, derived from it a consequence that he regarded as a *reductio ad absurdum* of the whole idea. Namely, were it really true that MZ twins are so similar with respect to IQ *solely* because they look similar and are for that reason treated similarly by others, then from the assumption that their phenotypic correlation is around 0.9 "it would follow that if only we could pass a child off as an MZ co-twin, perhaps with a few falsehoods and some alteration to his appearance, we could determine his IQ to within about 5 points. This trick should, for example, raise the IQ of a child otherwise doomed to one of 50, say, to 100 or even 150 depending on with whom he is paired" (Fulker 1975: 517; cf. Lykken 1995: 78).

3.2.5 The "attractiveness-IQ" scenario

To see why Fulker's skepticism is entirely justified let us consider the whole issue more carefully. Clearly, if a phenotypic correlation between monozygotic twins reared apart is very strong, one possible explanation is that the trait in question has high heritability. In principle, of course, other explanations are also possible. But being just *possible* is not enough to merit our attention. An additional requirement is minimal plausibility: we should not take seriously a scenario that is logically possible *and* in all probability false. The surprising thing about some of the repeatedly proposed "rivals" to the high heritability hypothesis is that they can often be dismissed as implausible to such a degree that there may be no need to look any further into empirical data that are relevant to the issue at hand.

Take a speculation about a connection between attractiveness and IQ, which is widely thought to put into doubt the inference to the heritability of IQ:

> Suppose, for example, that a child's perceived attractiveness and self-confidence strongly affect how adults interact with children *in a way that largely accounts for the variation in IQ* . . . Suppose further that personal attractiveness and self-confidence are highly heritable. Then we would have an indirect effect *par excellence*, and such an effect could, for all we know, *largely* account for the heritability of IQ. (Block 1995: 119 – italics added)

> Let us return to the speculation mentioned above that the 60 percent heritability of IQ . . . is *entirely indirect* and due to differential treatment of children on the basis of heritable characteristics. The direct heritability of IQ would be *zero*. (Block 1995: 121 – italics added)

Block's speculation is certainly a possibility (i.e., it is not self-contradictory). But once we scrutinize his scenario more closely and trace some of its implications it becomes clear that it is a non-starter.

Figure 3.1a gives the basic facts of the situation. The IQ correlation between MZA twins in adulthood is 0.75[11] (McGue & Bouchard 1998: 5), and the correlation between their genotypes is 1 (they have the same genotype). This much is assumed to be accepted and uncontroversial. However, the strengths of other causal arrows in the diagram are unknown. They are yet to be determined, and different explanatory hypotheses ascribe different values to them.

Consider two contrary accounts of the MZA correlation: (1) it is entirely explained by direct genetic causation, or (2) it is entirely explained by indirect genetic causation (G–E correlation). Let us see what the consequences of these two accounts are, and how they reflect on their plausibility.

---

[11] This is a weighted average of five studies of monozygotic twins reared apart.

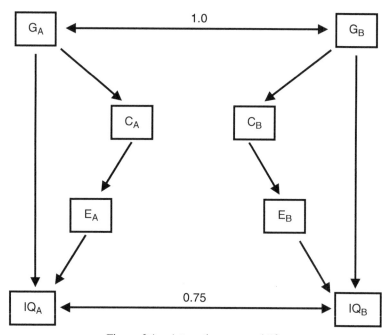

Figure 3.1a Attractiveness and IQ.

Referring to Figure 3.1b, model (1) with no indirect causation (i.e., without G–E correlation) would assign the value of 0.86 to two direct causal paths from G to IQ, while setting the value of the combined indirect path G→C→E→IQ to zero (thereby making all these paths disappear from the picture). This way, the phenotypic correlation between the twins would come out right (0.75). Namely, by the rules of path analysis, this correlation equals the product of all the paths connecting the two phenotypes: $0.86 \times 1 \times 0.86 = 0.75$. This model may or may not correspond to reality, but at least there is nothing about it that makes it intrinsically untenable or *a priori* unbelievable.

The situation is very different with model 2 (no direct causation, i.e. zero heritability). Now under the assumption that the *entire* phenotypic correlation is the result of *indirect* genetic effects, direct G-P arrows disappear from the picture. As shown in Figure 3.1c, we again use the same rule (the twins' IQ correlation equals the product of all the paths connecting their IQs), but this time, because of the indirect causation, there are *six* causal paths[12] (instead of just two, as in Figure 3.1b). It is easy to see that,

---

[12] Not counting the twins' genotypic correlation (1.0), which obviously cannot influence the final result when paths are multiplied with one another.

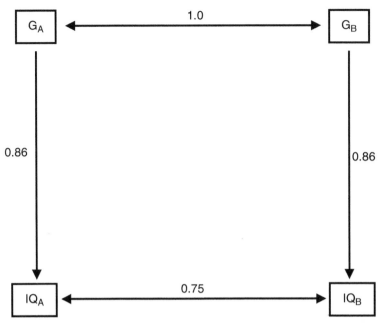

Figure 3.1b Heritability without G–E correlation

arithmetically, the product of the six paths (each of which is smaller than 1) can yield 0.75 only if *all* of them are *extremely* strong. Even setting each of these correlations to be very close to 1 (or, to be precise, 0.95) is still not quite enough to give the required value of the IQ correlation (0.75), although it very nearly approaches it.

But the correlations of that magnitude (0.95) between these variables are totally unrealistic. One indication that there is something fundamentally wrong with such an assumption is that the reliability of IQ tests (i.e., test–retest correlation) is known to be around 0.9. Therefore, it is out of the question that the correlation of IQ with something else could be stronger than its correlation with itself. Besides, the presence of correlations of the order of 0.9 between the three observable variables (attractiveness, particular environmental influence, and IQ) would be very noticeable and easily detectable.

However, far from this being the case, empirical research in fact shows that these correlations are not just weak but virtually non-existent. The first study showing that physical resemblance and IQ are not correlated was published in 1932 (Burks & Tolman 1932), thus undermining the explanatory scenario that Ned Block proposed in all seriousness more than sixty years later! Subsequent research gave similar results. In a large

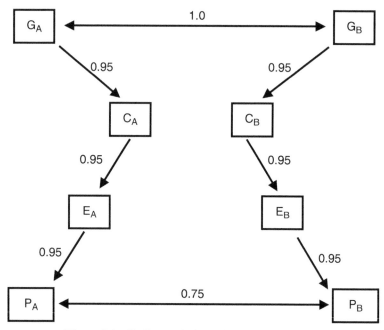

Figure 3.1c G–E correlation with zero heritability.

sample (N = 3,497), the median and mean correlation between attractiveness and IQ were found to be 0.00 and −0.04, respectively (Feingold 1992). This finding was confirmed for adult subjects in Langlois et al. 2000. Even those who argue that attractiveness and IQ are statistically associated (Kanazawa & Kovar 2004) interpret the correlation between the two variables as being non-causal (i.e. as resulting for assortative mating).

Another difficulty for the proposed indirect genetic effects on psychological differences is that many of the usually conjectured environmental influences may be empirically ruled out on general grounds. For instance, the idea that a given genotypically mediated trait makes the subject exposed to increased cognitive stimulation by others, which then in turn leads to the subject's increase in IQ, is contingent on the assumption that intellectually stimulating environments do have a substantial impact on IQ. However, although different families obviously differ greatly from one another with respect to the degree of intellectual stimulation they offer to the children, this between-family variation appears to account for *no* part of IQ variance among adults.[13] In other words, two genetically

---

[13] The worst environments are not represented in adoption studies, so the conclusion should be qualified. Yet it seems to hold for most environmental differences.

unrelated persons who are raised in the same family show on average as much similarity with respect to IQ in adulthood as an arbitrarily chosen pair of individuals. Hence, given that most of the systematic environmental influences hypothesized in the genetic indirect causation scenarios vary between families, the generally recognized lack of impact of these factors on IQ must make it additionally difficult to take these speculative scenarios seriously.

3.3 THE SOCIOLOGIST'S FALLACY

In this chapter genotype–environment correlation was considered as a problem for hereditarianism. Our discussion would not be complete, however, if we did not address a similar problem for environmentalism.

We saw that, when confronted with a correlation between G (genotype) and P (phenotype), wise hereditarians do not immediately jump to the conclusion that G caused P (G → P). They allow for the possibility that the true causal story may be G → E → P, with E being explanatorily much more important than G (and the genetic "first" cause even being de-emphasized in heritability estimates). But then wise environmentalists should be cautious as well. When discovering a correlation between E and P they should check for the possibility that E and P are not causally connected at all, and that their correlation is the result of E and P just being separate effects of G.

From the perspective of general causal analysis, the environmentalist's mistake is more serious because the danger here is to mistake a *spurious* cause for a *real* one, whereas in the hereditarian case the danger is to mistake an *indirect* cause for a *direct* one.[14] But in fact the environmentalist mistake is so frequent that Arthur Jensen honored it with a name, "the sociologist's fallacy." Among sociologists, in particular, there is a tendency to interpret the correlation between a social variable and phenotype as a causal relation, without even considering the possibility that genetic influences might be behind the correlation, making it completely bogus.

Of course, mistakes in dealing with statistical correlations are the reality of social science. In both orientations (hereditarian or

---

[14] "Ordinarily we would consider a failure to control for intervening variables as a less serious error than shortcomings in the control for prior variables. Clearly, the distinction between spurious and causal effect is of greater importance than the distinction between the two kinds of causal effect" (Hellevik 1984: 27–28).

environmentalist), there are researchers who err on the opposite side of caution and who hastily and fallaciously draw causal conclusions from statistical data. What is unexpected and worth mentioning, however, is that the sociologist's fallacy seems to be committed by a number of scholars who specifically address methodological issues and whose very goal is to critically examine the logic of different arguments in the debate. Here are several examples.

In the report on genetics and human behavior issued by the prestigious Nuffield Council on Bioethics, the genetic explanation of racial differences in IQ is mentioned and then accompanied by the following comment:

> While it is often claimed that on average, black individuals score slightly lower on IQ tests than white individuals, who in turn score lower than people from East Asia, there are also studies which show that, if black individuals and white individuals are closely matched on socioeconomic status, the differences in IQ are substantially reduced . . . The authors [of a study cited in the text] conclude that "socioeconomic differences are largely responsible for the usually reported differences in intellectual performance." (Nuffield 2002: 69)

The inference to the truth of the environmental explanation is premature here because the genetic hypothesis can also account for the decrease in the IQ difference between the two groups in this situation. For, it may be that matching whites and blacks on SES has the consequence that the two groups are matched on genetic characteristics as well. If that is the case, it may be that it is the genetic similarity that causes IQ similarity. Without more evidence there is no way to choose between the two possible explanatory hypotheses.

In a special supplement to the renowned Hastings Center Report, Erik Parens sets out to give a detailed methodological evaluation of research in behavior genetics. At one point he writes:

> Some very small (and old) studies have attempted to eliminate "environmental" differences by considering only blacks and whites raised in similar socio-economic circumstances. The results are mixed, some suggesting that the black–white test-score difference persists and some suggesting that it disappears . . . Those studies (by friends and foes of the genetic explanation for the gap) assume, however, that if we compare blacks and whites from the same "class," the environmental factors that might distort the comparison will be eliminated. That assumption could be true only if racism no

longer has effects. And there are strong reasons to reject that assumption. (Parens 2004: S15)

Again, the idea is to look for a cause of the white–black IQ differ-ence by considering blacks and whites matched on SES. Parens is right that the general environmentalist hypothesis is not refuted if the IQ dif-ference persists, because environmental influences *other* than SES might be responsible for it. But it is a curious omission that, speaking about what these studies assume, he does not mention a questionable assump-tion on which the refutation of the genetic explanation crucially depends, namely, that matching the individuals from the two groups on SES does not also match them on trait-relevant genetic characteristics. In other words, Parens points out, correctly, that the persistence of the IQ differ-ence (after controlling for SES) would not automatically refute the envi-ronmentalist hypothesis, but he fails to point out that the disappearance of the IQ difference (after controlling for SES) would not automatically refute the genetic hypothesis either. This asymmetry of treatment has no logical justification.

John M. Thoday, a highly esteemed geneticist and Ronald A. Fisher's successor as professor at the University of Cambridge, committed the sociologist's fallacy too: "The Negro population is, of course, low in mean socioeconomic status (SES), so that it is reasonable to postulate this as an environmental factor in accounting for some of the group difference" (Thoday 1976: 149). Of course, the SES difference *might* be a part of the explanation of the group difference in IQ but this *cannot* be inferred merely from the fact that the groups differ on SES. The third possible variable, genetic influence (G), has to be controlled for.

It is interesting that in these three examples methodologically sophis-ticated scholars made the environmentalist hypothesis look deceptively plausible by overlooking the need to test a crucial assumption, and that in each case this happened when they discussed the explanation of racial differences in IQ. Can it be that this fallacious reasoning with a strong environmentalist bias was unconsciously content-driven?

In more extreme cases, a reasonable warning that one should pro-ceed cautiously and consider the possibility that the correlation of non-biological variables could be spurious (because produced by the genetic common cause) is derided as obsession with genetics. So when Lindon Eaves set out to check whether the correlation between traumatic life events and depression might in part be explained by their heritability, Ruth Hubbard (the Harvard professor of biology and philosopher of biology) called this kind of research "ridiculous" and asked: "But once he [Eaves] found a clear correlation between traumatic events and depression, why look for a genetic explanation?" (Hubbard & Wald 1993: 6). Well, the answer to the question is: however "clear" a correlation may be, it never carries its causal truth on its sleeve. A genetic explanation has to be considered simply because it is one of the possibilities. In fact, further research has empirically confirmed that "ridiculous" hypothesis (Rijsdijk et al. 2001).

Another example of the sociologist's fallacy is the interpretation of the 2002 study that found a gene-dependent correlation between parental abuse and antisocial behavior. The gene in question is responsible for producing the neurotransmitter-metabolizing enzyme, monoamine oxidase A, called MAOA. According to Caspi et al. 2002, people who have a high activity variant of the MAOA gene do not manifest an increased level of antisocial behavior, whether they were abused in childhood or not. But those with the low activity MAOA gene behave antisocially significantly more often, but *only* if they were *also* abused by their parents. The authors conclude that the data constitute "epidemiological evidence that genotypes can moderate children's sensitivity to environmental insults" (Caspi et al. 2002: 851).

But this is jumping to the conclusion. The inference that parental abuse is a causal factor is premature. The fact is that those low MAOA subjects who were abused in childhood received two things from their parents: abuse *and* genes. Without further research we are simply not in a position to say which of the two factors was causally operative. It might be that antisocial behavior in the low MAOA group was caused by abuse, but then again it might be that the subjects inherited *other* genes from their (abusive) parents that predispose toward antisocial behavior. The hypothesis of gene–environment interaction (defended by the authors) has no epistemological advantage over the alternative hypothesis of gene–gene interaction.

Discussing the same topic, David Wasserman (2004: 27) also talks about "psychological effects of maltreatment" as something established, not something to be tested. Both Wasserman and the authors of the original study proposed several research designs for future studies that would test alternative explanations but they did not even mention the possibility of a purely genetic effect. Similarly, it has been claimed that the study of Caspi et al. "*demonstrates* the connection between genetic and environmental influences on behavior" (Nuffield 2002: 95 – italics added), that it found "the interaction of childhood abuse and a gene" (Parens 2004: S8), that the study "makes clear that a 'bad' genotype is not a sentence; for ill effects to occur, a bad environment is also required" (Ridley 2003b; cf. Ridley 2003a: 268).

All these authors commit the sociologist's fallacy. They leap to the conclusion about the causal impact of an environmental influence just on the basis of a familial correlation, without even considering the possibility that the correlation between abuse and antisocial behavior could be mediated by genes transmitted from parents to children. Matt Ridley's treatment of this topic is particularly odd. He mentions in a footnote (Ridley 2003a: 310) that he agrees with Judith Harris, who warned him that the conclusion about the environmental effect is unjustified, yet he kept drawing exactly that conclusion in the main text.

The strange thing about the sociologist's fallacy is that it is so widespread, despite many scholars issuing warnings about it for decades. In an important paper from thirty-five years ago Paul Meehl explained it very clearly and called it "the commonest error" (Meehl 1970: 394). To make things worse, Meehl said there that the mistaken reasoning had been clearly explained and criticized already in 1928 (Burks & Kelley 1928), but he stressed that it was still necessary to repeat the point. In desperation he quoted André Gide's remark: "It has all been said before, but you must say it again, since nobody listens." Unfortunately, judging by the quotations given here, Gide's remark is still relevant even today.

Another form of the sociologist's fallacy consists in an uncritical preference for environmental explanations, rather than in total neglect of genetic alternatives. It happens sometimes that before empirical evidence is evaluated at all, the environmental account is tendentiously presented as highly plausible, whereas the genetic explanation is allowed to appear only in a caricatured form that makes it look like an absolute non-starter. A good example is Kitcher's way of introducing two rival explanations of violent behavior:

> unless we are profoundly deceived, there are some readily identifiable features of the physical and social environment that have major impact: rates of crime are much higher in decaying inner cities, but I doubt that there is a "violence" allele that has the pleiotropic effect of sending its bearers into grim urban environments. (Kitcher 2001b: 409)

Well, yes, the scenario in which a segment of DNA dispatches its bearers into particular city areas does sound quite ridiculous. But, surely, there are other more complex, and less silly, ways the genetic causes might explain at least a part of the correlation of violent behavior and living in "grim urban environments." The possibility of genetic involvement cannot be dismissed by considering the most preposterous hypothesis of this kind. Deriding a genetic story that no one takes seriously will not turn a correlation between an environmental variable and phenotypic characteristic into a causal relationship.

Besides, Kitcher's approach is unprincipled because his methodological stance on causal inference changes radically, depending on whether he is talking about genes or environments. In the former case (see 3.2.1), he insisted that the genetic-cum-phenotypic similarity (in twin studies) was insufficient evidence for genetic causation, *even though a lot of empirical research in behavior genetics did undermine the environmental mediation hypothesis*. In the latter case, however, Kitcher is rushing to infer environmental causation of violent behavior from the environmental-cum-phenotypic similarity, mentioning only dismissively the possibility of genetic mediation. This is especially strange coming from a philosopher who defended the symmetry of genetic and environmental influences. Whatever happened to "causal democracy"?

4

From individuals to groups: genetics and race

> Enough of arguments of principle. In this case [the black–white difference in IQ], above all, they should be treated with grave suspicion. If it is easy enough to select data to suit one's prejudices, how much easier will it be to choose the arguments of principle which prove or disprove on *a priori* grounds that which one wished to conclude on other grounds?
>
> N. J. Mackintosh

The main source of political nervousness in discussions about heritability is its possible implications for race differences. John Searle echoed a widespread view: "once you believe that there are innate human mental structures it is only a short step to argue that the innate mental structures differ from one race to another" (Searle 1976).

The fact that the step from individual heritability to group heritability is perceived as "short" may explain the occasionally acrimonious opposition to claims about individual heritability: better to stop the inference at an early stage than to find oneself in the position where later, after conceding too much, one no longer has a good strategy to resist the abominable conclusion. But whatever resistance could be expected at the level of general discussions about heritability, the situation dramatically changes when race differences are addressed explicitly. Linda Gottfredson conveys the mood well: "One can feel the gradient of collective alarm and disapproval like a deepening chill as one approaches the forbidden area" (Gottfredson 1994: 56).

The most alarming hypothesis, of course, is the explosive mix of three ideas: race, IQ, and heritability. The hypothesis that racial differences in IQ are heritable has been attacked from all three directions. It has been claimed (1) that the concept of race is biologically meaningless, (2) that IQ

127

tests are biased to the point of being useless, or that they do not measure anything socially important, and (3) that there is no evidence that any psychological differences between groups are heritable.

Clearly, this is not the place to argue about (1) and (2), although I am convinced that they are both false. As for (3), comprehensive discussion is also out of the question. Rather, I will focus on only one aspect of the debate. Here again there has been a massive attempt to settle the controversial issue about within-group and between-group heritability by a knock-down *methodological* argument. I will try to show that this argument contains a colossal distortion of hereditarianism about race differences in psychology, and that it is astonishing that such a sham refutation has been accepted so widely and so long.

4.1 THE "MASTER ARGUMENT"

Many philosophers of science think that there is one particular argument that pinpoints the fundamental weakness in the proposed genetic explanation of the black–white IQ difference.[1] The argument says that the proponents of the genetic hypothesis about the inter-racial gap in IQ arrive at their conclusion by using a blatantly fallacious inference. The suggestion is that Arthur Jensen and the authors of *The Bell Curve* believe (wrongly) that from the claim that IQ is highly heritable among whites and among blacks it immediately follows that the difference in IQ between whites and blacks is also heritable. Once the mistake is diagnosed, we are told, the genetic hypothesis loses all its credibility. Here are some characteristic quotations:

> the existence of significant heritability for IQ within the populations that have been studied does *not* imply that average IQ differences between races are in whole or in any part due to genetic differences . . . Various writers – the most prominent being Arthur Jensen . . . – have taken the heritability of IQ to show that these differences must have a genetic base. No such conclusions follow. (Papineau 1982: 97)

> On the basis of evidence supporting a high heritability value within a subpopulation, Jensen infers that heritability will be (correspondingly?) high in the population as a whole, and that variation between groups has a

---

[1] It has been a consistent result of measurements that the difference between average IQ scores of whites and blacks in the United States is around 15 points. The controversy is about implications of this fact, and even more about its explanation.

(correspondingly?) high genetic basis . . . But there is no intrinsic connection between the magnitude of the heritability within groups and the
magnitude of between-group differences. (Richardson 1984: 401, 406)

Arthur Jensen's 1969 article in the *Harvard Educational Review* [Jensen
1969b] started off a current controversy by arguing from heritability within
whites to genetic differences between whites and blacks. In 1970 Richard
Lewontin gave a graphic example that illustrates why this is a mistake.
(Block 1995: 110)

The [hereditarian] argument is based on the assumption that if IQ has high
[heritability] in two different populations, then it can be concluded that the
difference in mean IQ between the two populations also has a high group
heritability. (Sarkar 1998: 93)

(See also Mazur & Robertson 1972: 89; Daniels 1976: 143–144, 173–174;
Rutter & Plomin 1997: 210; Nuffield 2002: 21.)

Helen Longino appears to subscribe to the same view because she finds
Richardson's analysis "persuasive" and "compelling" (Longino 1990: 8–
9). Since the same argument also occupies a central place in Block (1995),
and since Philip Kitcher describes Block's article as "the single best diagnosis of the flaws of Richard Herrnstein and Charles Murray's *The Bell
Curve*" (Kitcher 1998: 51; cf. Kitcher 1999: 89), it is fair to include Kitcher
among those who endorse it.[2] John Dupré also seems to support the view
because he says that "the confusions in this area are particularly clearly
explained by Block (1995)" (Dupré 2001: 30), and he calls Block's paper "a
powerful critique of the argument [of *The Bell Curve*]" (Dupré 2003: 129).
Among psychologists, Richard Nisbett thinks that Block offers "a particularly lucid account of heritability" (Nisbett 1998: 87). (Others who quote
Block's article with approval include Sternberg & Grigorenko 1999: 536,
538; Godfrey-Smith 2000: 27; Godfrey-Smith 1999: 330; Oyama 2000b:
S339; Gilbert 2002: 130; Rutter 1997: 391; Moore 2001: 45, 250; Ariew
1996: S21; De Jong 2000: 617; Kaplan 2000: 27, 69–70, 72; Parens 2004:
S14; Spencer & Harpalani 2004: 56; Vitzthum 2003: 546.)

The simple argument that has impressed philosophers of science so
much[3] was first used by Richard Lewontin in his criticism of Jensen

---

[2] Here is another consideration showing that the inclusion of Kitcher is justified: Rose,
Lewontin, and Kamin (Rose et al. 1984: 117–118) use the same argument against Jensen,
and Kitcher describes their chapter on intelligence as "brilliant," "lucid," "thorough," and
"devastating" (Kitcher 1984: 9).

[3] As a referee for one of my papers published in *Philosophy of Science* said: "This argument
has become practically canonical in our profession."

(Lewontin 1976c). Ever since, Jensen's views have been routinely dismissed by rehearsing Lewontin's well-known example of two handfuls of seed taken from the same, genetically heterogeneous sample and then planted in two different soils (rich and poor in nutrients): as a result, the phenotypic differences *within* each of the two groups of plants will be 100 percent heritable, but the difference *between* the two groups will be entirely due to differences in two environments (zero heritability). The moral triumphantly drawn from that example is: "You just cannot establish *between*-group heritability on the basis of *within*-group heritability!" Indeed, this is correct. The only question is whether Jensen was really unaware of that elementary truth.[4]

Richardson for one found it amazing that Jensen could have been so "blind."[5] And he offered the following explanation:

> How might we explain this blindness on Jensen's part? It is exactly here that the point that his doctrine is a racist doctrine – as it manifestly is – enters in. The latent racism explains the persistence of the view despite its manifest untenability on scientific grounds. (Richardson 1984: 407)

It is regrettable that philosophers are so keen to resort to the heavy artillery of political accusations. If it *seems* that a respected scholar has been "blind" about something very simple and elementary, minimal fairness requires that you think twice (and read twice) in order to make sure that perhaps it was not you who misunderstood the person in question. I will try to show that in this particular case Richardson jumped the gun, and that a closer look at the texts resolves the problem in a way that does not necessitate a speculation about Jensen's political motives. More generally, I will claim that Lewontin's "master argument" is really a red herring that has directed the philosophical discussion away from the real issues for the last thirty-five years.

---

[4] Elliott Sober also uses Lewontin's example against *The Bell Curve*, and before introducing it he says: "The point I want to make here is that [Herrnstein and Murray's] conclusion [that the observed difference between white and black Americans in IQ has a genetic basis] would not follow even if variation within the two groups had a significant genetic component" (Sober 2000: 362). Sober became more cautious in discussing Lewontin's criticism of Jensen in Sober 2001: 58.

[5] Louise Antony goes even further than that. Discussing *The Bell Curve*, she claims that Herrnstein "knows perfectly well" that the high within-population heritability cannot be used to support the genetic explanation of a between-population difference ("because it has long been pointed out to him by critics like Harvard geneticist Richard Lewontin"), but she says that Herrnstein nevertheless "willfully" continued committing the fallacy (Antony 1997: 482).

Let me start by describing three possible ways of arguing from WGH (within-group heritability) to BGH (between-group heritability).

(H_1) High WGH entails a non-zero BGH.

(H_2) High WGH, by itself, inductively establishes a non-zero BGH.

(H_3) High WGH, together with some collateral empirical information, inductively establishes a non-zero BGH.

While serious hereditarianism actually involves commitment to H_3, Lewontin and the philosophers of science following in his footsteps have persistently criticized H_1 (or occasionally H_2), with the unfortunate result that they simply never managed to get in contact with the real hereditarian argument (which aims to support H_3).

When Lewontin says, for example, that "[t]he fundamental error of Jensen's argument is to *confuse* heritability of a character within a population with heritability of the difference between two populations" (Lewontin 1976c: 89 – italics added), or that "[t]here is, in fact, no valid [*sic*] way to reason from [WGH] to [BGH]" (Rose et al. 1984: 118), or that he was "taken aback by the obvious and elementary conceptual and logical errors in this argument" (Lewontin 1976d: 97), this only makes sense as an attack on H_1. Again, his assertion that "[t]he genetic basis of the difference between two populations bears no logical or empirical relation to the heritability within populations and *cannot be inferred* from it" (Lewontin 1976c: 89 – italics added) is best understood as a criticism of H_1, or perhaps H_2. Furthermore, accusing Jensen of the "manifestly incorrect claim" (Lewontin 1975: 399) and "elementary error" (402) sounds more like suggesting an outrageous, inexcusable blunder (like H_1) than like disputing a fairly complex, empirically based bit of reasoning (like H_3), where there might be room for disagreement among reasonable and competent people.

Gould also seems to attribute H_1 to Jensen because he speaks of the "confusion" of within- and between-group variation and "the *non sequitur* of the worst possible kind," pointing out that there is "no necessary relationship" between WGH and the difference between the two group means (Gould 1977: 246). Finally, and most importantly, the very fact that Lewontin thinks that he can disprove a connection between WGH and BGH by merely using his famous example of two handfuls of seed (Lewontin 1976c: 89; Rose et al. 1984: 118; see also Block 1995: 110) shows that the target here cannot be H_3. For, although his thought experiment does persuasively demonstrate the logical fallaciousness of any attempt to immediately *derive* non-zero BGH from high WGH (*à la*

H_1), it is by itself irrelevant for the evaluation of a much more sophisticated hypothesis[6] (embodied in H_3).

If any more evidence is needed, the following quotation should dispel all doubts about the severity of the mistake that Lewontin attributed to Jensen:

> The error of confusing the heritability within a population with the causes of differences between populations was clearly made by Arthur Jensen in his famous article in the *Harvard Educational Review*, when he tried to infer from heritability studies within the American white population the causes of differences between races. This elementary blunder would not be tolerated in a freshman class in statistics or genetics. We may well wonder how it came to be made by a professor! (Lewontin 1973)

The bad news for the critics inspired by Lewontin's conceptual line of attack on hereditarianism is that Jensen in fact never intended to defend H_1 (or H_2). I will first support this interpretation by quoting from Jensen's publications. Later, I will exhibit the logical structure of his real argument (that happens to be a version of H_3).

The first indication that Lewontin's argument may have missed the mark badly is Jensen's reaction in his exchange with Lewontin:

> The main thrust of Lewontin's argument, as he sees it, actually attacks only a straw man set up by himself: the notion that heritability of a trait within a population does not prove that genetic factors are involved in the mean difference between two different populations on the same trait. I agree. But nowhere in my *Harvard Educational Review* discussion of race differences do I propose this line of reasoning, nor have I done so in any other writings. (Jensen 1976b: 103)

Of course, we don't have to take Jensen's word here, but if an author protests that his views are distorted this certainly constitutes a good reason to proceed cautiously and explore the matter in more detail. Besides, since the fact that "there is no simple correspondence between the contributions of nature and nurture to group and individual differences" was treated as common knowledge already thirty years before the publication

---

[6] In fairness to Lewontin, it should be pointed out that, besides discussing heritability, he does also briefly address the question whether "the lack of effect of correction for gross socioeconomic class" might be a "presumptive evidence" for hereditarianism (1976c: 91). But in this short discussion he nowhere shows that he recognizes how Jensen harnesses global empirical evidence to work *together* with high WGH to build the hereditarian case (see below). For Jensen, *both* the relative weakness of environmental influences on IQ *and* high WGH are *essential* parts of the inference to non-zero BGH. Disputing these two premises separately (as Lewontin does) is an *ignoratio elenchi*.

of Jensen's paper (Burks 1938: 277), it sounds odd to suggest that Jensen was not aware of it. Another sign that something has gone awry with Lewontin's understanding of Jensen is the following comment of the geneticist James Crow in the earliest round of comments on Jensen's 1969b paper: "Strictly, *as Jensen mentions*, there is no carryover from within-population studies to between-population conclusions" (Crow 1969: 159 – italics added). So, what Lewontin accuses Jensen of not seeing is the same thing that in Crow's opinion Jensen not only saw but *mentioned* as well!

The best way to resolve this odd disagreement is to go directly to what Jensen actually wrote:

> So all we are left with are *various lines of evidence, no one of which is definitive alone, but which, viewed all together,* make it a not unreasonable hypothesis that genetic factors are strongly implicated in the average Negro–white intelligence difference. (Jensen 1969b: 82 – italics added)

This statement[7] is manifestly incompatible with the belief (attributed to Jensen by many of his critics) that within-group heritability *alone* is sufficient to establish the between-group heritability. Indeed, immediately after mentioning "various lines of evidence," Jensen clarifies his position further by summarizing *six* different empirical arguments that, in his opinion, *together with the high within-group heritability of IQ*, lend support to the genetic explanation of the between-group difference.

Addressing the relation between within-group heritability and between-group heritability, Jensen made essentially the same point in many of his other writings. For example:

> I have explained in greater detail elsewhere [1968] that *heritability coefficients by themselves cannot answer the question of genetic differences between groups*, but when used along with additional information concerning the amount of relevant environmental variations within groups and overlap between groups, can enter into the formulation of testable hypotheses that could reduce the heredity–environment uncertainty concerning group differences. (Jensen 1969a: 220 – italics added)

> other methods than heritability analysis are required to test the hypothesis that racial group differences in a given trait involve genetic factors and to determine their extent. (Jensen 1973b: 411)

---

[7] Interestingly, Lewontin quotes a section containing that very statement, but he refuses to take it at face value and says that this "cant" needs to be "translated into common English" (Lewontin 1976c: 89). In the "translation" he then proposes, the crucial part of Jensen's statement is lost.

> Although one cannot formally generalize from within-group heritability to between-groups heritability, the evidence from studies of within-group heritability does, in fact, *impose severe constraints* on some of the most popular environmental theories of the existing racial and social class differences in educational performance. (Jensen 1973a: 1 – italics added)

(See also: Jensen 1968: 95; 1973a: 135–139; 1977: 232; 1981b: 502–504; 1982: 126; 1994: 905; 1998: 445–463.)

In mainstream philosophy of science, Lewontin's argument against Jensen is repeated ad nauseam, but Jensen's response, if mentioned at all, is dismissed without being properly explained, let alone evaluated or critically considered. To the best of my knowledge, in no other context have philosophers of science demonstrated a systematic bias of such dimensions in presenting the ongoing scientific debate.

The most surprising thing is that the charge against Jensen (that he tried to infer BGH immediately from WGH) is routinely made and then readily transmitted further, without anyone feeling an obligation to produce textual evidence for that attribution. And when, exceptionally, someone does attempt to provide a supporting quotation it turns out that he is actually unable to deliver:

> Does high IQ heritability in the white population, combined with a 15 point black–white mean difference, permit us to conclude anything about the reasons for, or causes of, the IQ gap? Jensen (Jensen 1972a: 163) clearly believes it does: "It is not an unreasonable hypothesis that genetic factors are strongly implicated in the average Negro–white intelligence differences." (Daniels 1976: 173)

Contrary to what is being suggested here, the quotation from Jensen only confirms that he advocates the non-zero BGH hypothesis; it certainly does *not* show that he inferred it directly from WGH (as Daniels implies).

One of the most outspoken critics of heritability, Marcus Feldman, went much further. In a court deposition as an expert witness for the defense in *Grutter v. Bollinger* (the case regarding the University of Michigan Law School's admissions policy), Feldman wrote:

> In Jensen's case and in the view of many genetically uninformed authors "the fact that intelligence variation has a large genetic component … makes it a not unreasonable hypothesis that genetic factors are strongly implicated in the average Negro-white intelligence difference." (Feldman 2001: 168)

The text in quotation marks is indeed Jensen's and it does look as though it contains a leap from WGH to BGH. But the problem is that, in Jensen's paper, the two sentence fragments separated by the ellipsis belong not only to different sentences but even to different paragraphs. If we focus on Jensen's authentic words, the "fallacy" vanishes and Jensen's claim becomes completely different from how Feldman represented it: "So all we are left with are *various lines of evidence, no one of which is definitive alone, but which, viewed all together,* make it a not unreasonable hypothesis that genetic factors are strongly implicated in the average Negro–white intelligence difference" (Jensen 1969b: 82 – italics added).

In my opinion, this kind of deliberate misrepresentation in attacks on hereditarianism is less frequent than sheer ignorance. But why is it that a number of people who publicly attack "Jensenism" are so poorly informed about Jensen's *real* views? Given the magnitude of their distortions and the ease with which these misinterpretations spread, one is alerted to the possibility that at least some of these anti-hereditarians did not get their information about hereditarianism first hand, from primary sources, but only indirectly, from the texts of unsympathetic and sometimes quite biased critics.[8] In this connection, it is interesting to note that several authors who strongly disagree with Jensen (Longino 1990; Bowler 1989; Allen 1990; Billings et al. 1992; McInerney 1996; Beckwith 1993; Kassim 2002) refer to his classic paper from 1969 by citing the volume of the *Harvard Educational Review* incorrectly as "33" (instead of "39"). What makes this mis-citation noteworthy is that the very same mistake is to be found in Gould's *Mismeasure of Man* (in both editions). Now the fact that Gould's idiosyncratic *lapsus calami* gets repeated in the later sources is either an extremely unlikely coincidence or else it reveals that these authors' references to Jensen's paper actually originate from their contact with Gould's text, not Jensen's.

It is worth stressing, however, that even those who did not read Jensen still had an alternative and very easy way to acquire better understanding of the position they dismissed so hastily. Merely consulting the writings of more sophisticated environmentalists and *opponents* of Jensen would have been quite sufficient to reveal the hollowness of Lewontin's "refutation." Surprisingly, this avenue has not been used either.

Among those currently defending the environmentalist explanation of the racial IQ gap the leading authority is James Flynn. At the beginning

---

[8] In more extreme cases, it happens that even some prominent academics publicly condemn hereditarian publications that, by their own admission, they have not read at all. For instance, the president of Rutgers University said in 1995 that he and his wife had both refused to read *The Bell Curve* because they considered its basic claim "morally wrong" (*New York Times*, February 6, 1995).

of his academic career Flynn was attracted to the debate by the desire to prove that Jensen was wrong, and indeed he ended up by providing some interesting arguments against hereditarianism. But despite disagreeing with Jensen, Flynn does not think that Jensen's view is the product of conceptual confusion, methodological fallacies, or racism. He takes pains to correct the widespread misinterpretations, and insists that Jensen never made the elementary mistake of inferring the genetic explanation of the between-group difference only on the basis of the high heritability of within-group differences:

> [Jensen] does not believe that [heritability] estimates alone can decide the issue of genetic versus environmental hypotheses. However, he argues that the probability of a genetic hypothesis will be much enhanced if, in addition to evidencing high [heritability], we find we can falsify literally every plausible environmental hypothesis one by one. He challenges social scientists who believe in an environmental explanation of the IQ gap between the races to bring their hypotheses forward. Given his competence and the present state of the social sciences, the result is something of a massacre ... Far too many of Jensen's critics have not taken up the challenge to refute him in any serious way, rather they have elected for various forms of escape, the most popular of which has been to seize on an argument put forward by the distinguished Harvard geneticist Richard C. Lewontin. (Flynn 1980: 40, 54)

Although Flynn's excellent book *Race, IQ and Jensen* came out in 1980 there is no reference to it in philosophical works addressing the same topic that were published later. This is a curious omission because Flynn's subtle methodological analysis of the whole debate is a kind of contribution that should be of special interest to philosophers of science. In particular, after Flynn's powerful and very detailed criticism of Lewontin's argument it is rather odd to see scholars defending that same argument with undiminished fervor and without any apparent awareness of the grave objections raised against it.[9]

Now, the best way to see why Lewontin's argument does not work is to be acquainted with the basics of Jensen's real stance. I will here briefly present an interpretation of Jensen's essential position that is both endorsed by Jensen himself and widely accepted by other scholars, hereditarians and environmentalists alike.

---

[9] Flynn later softened his attitude toward Lewontin (Flynn 1989: 365–366) mainly because he thought that he had himself discovered *empirical* evidence pointing to a possibly workable Lewontin-like scenario. This idea itself has serious problems (see Nichols 1987) but even if it were accepted, Flynn's basic argument against Lewontin's aprioristic attack on hereditarianism would still remain absolutely cogent.

The most important thing to recognize at the outset is that Jensen's inference proceeds in *two* steps. The first step consists in arguing for the substantial within-group heritability of IQ. But this is only the first step. Contrary to what Lewontin and his philosopher-followers suggest, Jensen in fact never claimed that this step by itself established (deductively or inductively) that the between-group differences are also heritable. Rather, Jensen thinks that high WGH puts severe *constraints* on admissible environmentalist explanations of the between-group differences in IQ. And then, *in the second step*, he argues on empirical grounds that, given these constraints, none of the proposed environmentalist hypotheses remains plausible.

Here is Jensen's argument in a schematic form.

(A) High WGH of IQ (among both whites and blacks).
(B) Empirical data (mainly about the relation of certain environmental variables and IQ).
(C) Non-zero BGH.

Critics charge that (C) cannot be inferred from (A). Granted. But making this point comes nowhere near to addressing the real hereditarian argument, which tries to reach (C) on the basis of *both* (A) and (B).

Another thing to keep in mind here: conceding that (C) cannot be inferred from (A) alone, deductively or probabilistically, should not be taken to mean (as it sometimes is)[10] that (A) is *evidentially irrelevant* for the truth of (C). Jensen's basic claim is that given (A), the empirical information in (B) makes (C) more plausible than (not-C). If (A) were false, however, he thinks that (B) would lose at least some of its force as a reason for accepting (C). So, in this picture, (A) is an essential argumentative part of the case for (C) although (A) is by itself insufficient to establish (C).

Admittedly, at several places Jensen does say that high WGH increases the probability of non-zero BGH (1976b: 104; 1973a: 135, 144; 1973b: 408), but on closer reading it becomes clear that he actually never commits himself to the claim that high WGH, by itself, inductively establishes non-zero

---

[10] Feldman and Lewontin write: "we are unable to make any inferences from between-group differences and within-group statistics about the degree of genetic determination of the between-group differences. In other words, the concept of heritability is of no value for the study of differences in measures of human behavioral characters between groups" (1975: 1167; cf. Feldman & Lewontin 1976: 13). They fallaciously conclude that the non-derivability of BGH from WGH means ("in other words") that WGH is of no value for any inference about BGH.

BGH. Here is how Jensen explains the reason for postulating the probabilistic relation between WGH and BGH: "In nature, characteristics that vary genetically *among* individuals within a population also generally vary genetically *between* different breeding populations of the same species" (1973b: 408; cf. 1969a: 80; 1972c: 974; 1973a: 130). He suggests that, as a rule (with almost no exceptions), high WGH of a trait is in nature accompanied by non-zero BGH, and he moreover states that this strong empirical association is regarded as a well-established fact by geneticists. These claims were never seriously disputed either by Lewontin or by philosophers of science opposing hereditarianism. However, Jensen warns explicitly that even if the general probabilistic relation between WGH and BGH is conceded, high WGH still does not allow any determinate conclusion about BGH with respect to a particular trait: *additional* empirical evidence is necessary. He concludes the relevant section with a cautionary remark: "As I have pointed out elsewhere, *other methods than heritability analysis* are required to test the hypothesis that racial group differences in a given trait involve genetic factors and to determine their extent" (Jensen 1973b: 411 – italics added).

Now moving forward from the schematic version, let us flesh out Jensen's argument. There are two different strategies for defending the environmentalist explanation of the racial IQ gap. Jensen tries to show that neither of these strategies looks plausible when high within-group heritability is *combined* with additional empirical evidence. The first strategy attempts to explain the difference between the two groups by invoking environmental factors that vary within the groups and that therefore enter into the calculation of WGH. Since these factors exhibit a within-group variation, I will call this type of environmentalist explanation "VE theory" (VE = variable environments). The second strategy tries to explain the between-group difference by postulating a factor that has no within-group variance but which is consistently present in one group and consistently absent in the other one. Following Jensen, I will label this kind of environmentalist explanation "X-factor theory." Let me first take up the VE theories.

4.2 VE THEORIES

Are VE theories automatically refuted by the mere fact of high WGH? It depends. If WGH is 100 percent the answer is yes. For, if WGH is 100 percent this means that differences in VE factors have no effect on

Table 4.1 *The rising constraints*

h^2	0%	10%	20%	30%	40%	50%	60%	70%	80%	90%	100%
VE	1	1.05	1.12	1.19	1.29	1.41	1.58	1.82	2.24	3.16	–

IQ variation within groups, and hence they cannot account for any part of the observed between-group difference in IQ either. On the other hand, if WGH is high but less than 100 percent then it is still possible that differences in (within-group) variable environments may explain the between-group difference in its entirety. This is in principle possible, but additional empirical evidence may rule out this possibility too.

High WGH tells us that most of the within-group variance in IQ is caused by genetic differences, and that environmental differences in VE factors are comparatively weak causes. Although they are not causally impotent, VE differences have relatively small effects. But then – and this is crucial for understanding the relevance of WGH – if VE differences are indeed so weak as causes the following is true: for the between-group difference to be fully explained by VE factors the two groups must mutually differ with respect to these VE factors *to a very large extent*. The difference in VE required for the complete environmentalist explanation can be so large that, in the light of available data about the group differences in VE, the very hypothesis will sometimes be seen as empirically hopeless. Table 4.1 shows the rising constraints that increased within-group heritability puts on pure between-group environmentalism.[11]

The values in the bottom row represent the number of standard deviations (of the distribution of the environmental component of IQ) required for explaining the total IQ gap.[12]

So, if heritability is 40 percent (or 60 percent or 80 percent respectively) the VE difference required to explain the whole IQ gap by environmental causes becomes 1.29 SD (or 1.58 or 2.24). The higher the heritability, the heavier the burden on purely environmentalist accounts. The gist of Jensen's position is precisely this two-barreled argument. He says, first,

[11] After having compiled this table I found out that Jensen gives a similar tabulation in 1998: 455. (There is an obvious misprint in his text, though: the required VE difference for the heritability of 40 percent should be 1.29 and not 1.77.) A nice graph used to make the same point can be found in Levin 1997: 126.

[12] These values are in each case obtained by dividing the entire inter-group IQ difference (15 points) by the respective standard deviation (SD) of the environmental component of IQ variance. The latter magnitude is inversely dependent on heritability, and is easily ascertainable once heritability is fixed.

that high WGH compels radical environmentalists to postulate very big VE differences between the groups and, secondly, that empirical observation shows that these enormous hypothesized differences are simply not there.

Although this is the barest outline of Jensen's argument I hope it shows how essentially his hereditarian conclusion rests on the empirical evaluation and eventual rejection of different versions of VE environmentalism. He carefully considers many prima facie promising attempts to account for the black–white IQ difference in terms of VE factors like SES, educational inequality, malnutrition, teacher expectations, childrearing practices, prenatal and perinatal disadvantages,[13] etc. (see, e.g., 1981a: 214–226; 1994: 905–906). But in the end he concludes that the explanatory burden imposed on these hypotheses by high heritability is too heavy and that these environmentalist hypotheses, singly and collectively, fall far short of explaining the total between-group IQ variance. From this brief exposition of Jensen's line of reasoning it should be obvious that the suggestion that Jensen derived BGH directly from a single premise (WGH) introduces a ridiculous caricature of hereditarianism that is as easy to refute as it is meaningless to discuss.

4.3 X-FACTOR THEORIES

What Lewontin's example of two batches of seed in different environments does show is that even in the case of 100 percent within-group heritability, the between-group heritability can still be zero. Bear in mind, however, that this can happen only if the phenotypic differences between the two groups are caused by an environmental factor that has no within-group variance at all. Hence, complete hereditarianism about *within*-group differences is logically compatible with complete environmentalism about *between*-group differences. All right. But Lewontin himself and too many of his ardent supporters thought that the example proved something much stronger, namely, that within-group heritability is *entirely irrelevant* for assessing between-group heritability. This is wrong. As Flynn says:

---

[13] Given Jensen's notorious efforts to take seriously every suggested environmental influence and discuss its empirical status in detail, it must come as a surprise that Ned Block could accuse him of basing his judgment on "*selected* facts" and on "excluding information about blacks' less favorable environments" (Block 1974).

the real message of Lewontin's example is that we can ignore high [heritability] only if there exists a highly specific and highly unusual set of circumstances. Therefore, it is absurd to say that high [heritability] estimates within black and white respectively are irrelevant. Their relevance consists precisely of this: they force us to look for a plausible candidate for the role of [X-factor]. (Flynn 1980: 58–59)

As both Flynn and Jensen point out, in reality it may be very difficult to find a plausible candidate for the role of X-factor. This is because this factor ought to be uniformly present in one group and uniformly absent in the other group, and furthermore it should manifest no variation inside either group. For this reason, SES and educational inequalities (the usual suspects in the puzzle of racial difference in IQ) are automatically excluded in this kind of scenario because they obviously have a significant variance within both whites and blacks. The same is true of some other popular candidates for an X-factor, and all this shows that the search will by no means be easy. Indeed, high within-group heritability can so severely constrain X-factor theorizing as to make this type of environmentalism exceptionally vulnerable to disconfirmation. This is exactly what Jensen was trying to demonstrate. And here again he certainly did not argue in favor of the genetic hypothesis by relying solely on the fact of high within-group heritability of IQ but by also extensively analyzing the empirical credentials of prospective X-factor hypotheses, and by finding them sorely wanting.

At first it might seem that there is actually an environmental factor that fits the bill smoothly: discrimination. The reasoning is straightforward: blacks, *as a group*, are exposed to pervasive discrimination and racism triggered by their skin color, whereas whites are never disadvantaged by the same kind of social prejudice targeted at their group. Therefore, the argument goes, since this environmental difference operates at the group level (exactly as an X-factor should), it forces itself upon us as a highly probable explanation of the inter-group difference in IQ. A great number of people have found that argument supremely convincing. But on second thoughts (if there happens to be a second thought, that is) the idea faces serious difficulties. Despite being initially quite plausible, the suggestion that discrimination is an easy answer to the racial IQ gap is flatly rejected by more sophisticated environmentalists as being "simply an escape from hard thinking and hard research":

But this is simply an escape from hard thinking and hard research. Racism is not some magic force that operates without a chain of causality. Racism

141

harms people because of its effects and when we list those effects, lack of confidence, low self-image, emasculation of the male, the welfare mother home, poverty, it seems absurd to claim that any one of them does not vary significantly within both black and white America. (Flynn 1980: 60)

Put differently, although discrimination at first looks like an X-factor, it may turn out that it can plausibly operate only through mechanisms involving a host of VE factors. This shows why thoughtful environmentalists don't rush to embrace the explanation by discrimination when invited by Jensen to try that route, and why instead they immediately sense danger and feel like being offered "a poisoned apple, an escape that looks attractive but proves fatal" (Flynn 1999b: 13).

Even if the X-factor does not exert its influence via VE factors, there is still a way to detect its presence. Assuming that it varies within the minority group (surely not all blacks are exposed to the *same* degree of discrimination!), the X-factor would increase the phenotypic variance in the affected group as well as the variance in any variable "touched" by the X-factor. David Rowe and his associates tried to uncover these signs of different developmental processes in whites and blacks but found none (Rowe et al. 1994; Rowe et al. 1995). This implication of increased variance is also a problem for the most recent attempt to explain the racial difference in IQ in purely environmental terms (Dickens & Flynn 2001), as pointed out by critics (Loehlin 2002; Rowe & Rodgers 2002). The issue remains empirically unresolved.

One should not conclude from all this that purely environmentalist scenarios (either of VE or X-factor variety) are ruled out of court. What the foregoing discussion was meant to show is just that high within-group heritability changes the terms of the debate in the sense that it puts additional obstacles in the path of pure environmentalism about group differences. Given the limited objectives of the approach in this book, no opinion is expressed about whether environmentalists can overcome these obstacles or not.

4.4 UNFAIR TO FACTS

The environmentalist criticism of hereditarianism that dominates contemporary philosophy of science is so crude and ill founded that it simply does not connect with the best discussions in the field. The issues are immensely more complicated than philosophers typically think. Ned

Block, the leading philosophical authority on the questions of race, IQ, and heritability, insists that the hereditarian position relies on "conceptual confusions" (Block 1995: 99), "flawed logic" (110), and "mistake" (ibid.). I have tried to show that, on the contrary, hereditarianism is a perfectly legitimate hypothesis: the alleged flawed logic and conceptual confusions are in this case just in the eye of the beholders (Lewontin, Block, Layzer, etc.). Properly interpreted, hereditarianism is a carefully argued and methodologically sound theory. Needless to say, this does not mean that it is true or that it should be accepted. The only reasonable way to take sides about this issue is to painstakingly examine the rich empirical material accumulated in the last several decades and to explore different lines of argument based on available data. There is no philosophical road to truth about these things. Knowledgeable environmentalists today are well aware that they cannot win the debate by just relying on the "intuitive plausibility" of their view, or by arguing that any other answer must be the product of muddled thinking or racism (or both).

I hope I have demonstrated in this chapter that Lewontin and philosophers who have followed him "refuted" Jensen by first distorting his position beyond recognition and that afterward all went quite effortlessly.

Ned Block does the same thing with *The Bell Curve*. He starts by attributing to Herrnstein and Murray the following principle (a version of WGH–BGH fallacy):

> if a characteristic is largely genetic and there is an observed difference between two groups, then there is "highly likely" ... to be a genetic difference between the two groups that goes in the same direction as the observed difference. (Block 1995: 102)

According to him, this principle "underlies all of Herrnstein's and Murray's thinking *even though it is never articulated*" (1995: 102 – italics added). This is odd. For it is not only that Block cannot provide textual evidence that the authors of *The Bell Curve* subscribe to the aforementioned principle. He does not give a single quotation that would at least show that they sometimes make an inference in accordance with that principle (which allegedly "underlies all of their thinking"). Even worse, Herrnstein and Murray in fact explicitly express a statement that goes *against* that principle, and they moreover put that statement in italics:[14]

---

[14] Curiously, Block quotes this statement in a footnote (and criticizes it for a reason we cannot go into), but he does not seem to notice a tension between that statement and the view he arbitrarily tries to impose on its authors.

> *That a trait is genetically transmitted in individuals does not mean that group differences in that trait are also genetic in origin.* (Herrnstein & Murray 1994: 298)

On the next page they say: "The heritability of individual differences in IQ does not necessarily mean that ethnic differences are also heritable." It is fairly clear that what Herrnstein and Murray want to say is that within-group information is by itself *insufficient* to establish a between-group conclusion. Additional information is necessary to make that step, and indeed a few pages later they adduce crucial *supplementary* evidence as the missing argumentative link, like the so-called "Spearman's hypothesis" and the weaknesses of extant environmentalist explanations (Herrnstein & Murray 1994: 301–311). One can of course dispute that inference too, but (as I have shown in Jensen's case) as soon as one acknowledges the essentially two-step structure of the argument, the debate must turn to empirical matters and one can no longer resort to the quick strategy of dismissing hereditarianism as just a crude methodological fallacy.[15]

It is really amazing how the factoid about "the big blunder of hereditarianism" (the mythical inference of between-group heritability from within-group heritability) is spread by otherwise serious scholars, without any feeling of obligation on their part to support the accusation with incriminating texts. Another example involves British psychologist Christopher Brand, whose book *The g Factor* (Brand 1996) appeared in the spring of 1996 but was abruptly withdrawn by the publisher soon after the first copies reached bookstores in the UK. (The book was never released in the USA.) Immediately after the book was "de-published," Steve Jones wrote an article in the British daily, the *Guardian*, in which he accused its author of the "elementary mistake" of inferring the between-group heritability from a statement about within-group inheritance (Jones 1996).

What made the whole situation particularly delicate was that, with the publisher's withdrawal of the book, the readers of the *Guardian* would have found it difficult to check for themselves whether Jones's presentation of Brand's views was fair. They probably assumed that it was, which seemed a reasonable assumption under the circumstances. But I read Brand's book very carefully (one copy found its way into a Hong Kong

---

[15] K. A. Appiah also attributes the WGH-to-BGH fallacy to Herrnstein and Murray, but again without providing any textual support. He says that the erroneous argument is "more implied than asserted" (Appiah 1995: 305). At another place, however, Appiah claims that the authors of *The Bell Curve* are aware that between-group heritability does not follow from within-group heritability (1996: 99).

library) and I have to report that I found absolutely no basis for Jones's imputation in Brand's text.

This is not the first time that Jones has "refuted" hereditarians by attributing to them non-existent errors. For example, here is how he summarizes the main argument of *The Bell Curve*: "Poor people (particularly poor black people) score lower on IQ tests than do rich; IQ scores run in families; class and race differences in IQ must therefore be due to genes" (Jones 1994b). The suggestion is that Herrnstein (a Harvard psychology professor) and Murray (a sophisticated social scientist) were not aware that a trait may run in families but be transmitted non-genetically. Now, surely, it is impossible that even Jones really believed that they did not know *that*. But apparently, the media were looking for a simple and blatant "*Bell Curve* Fallacy" with which to dismiss the book, and Jones was willing to deliver the goods.

An odd thing about Jones is that, despite his authoritative-sounding criticisms of human behavior genetics in newspaper articles and TV appearances, he actually has no single peer-reviewed publication that comes anywhere close to this area of research. As he himself once said: "Although I write a lot about it, I've never done any serious work of my own in human genetics" (in Brockman 1995: 118). One must wonder why the media nevertheless turn to him so often for an "expert" opinion on these issues.

In stark contrast with the empirical orientation and awareness of complexity of the issues that rules among the best advocates of environmentalism and hereditarianism, many scientists and philosophers of science still largely live in their own, socially constructed world. They are massively beguiled by the belief that hereditarianism can be easily checkmated in a couple of moves discovered by conceptual analysis. Mario Bunge suggests that hereditarianism about racial differences in intelligence is charlatanism, not science. He says that Jensen's hypothesis that the lower IQ of blacks is partly due to genetic factors "was unanimously rejected by the scientific community" (Bunge 1996: 106). In actuality, according to the poll of experts in the relevant fields, of all the scientists who felt qualified to express a view on that issue, 53 percent agreed with Jensen (Snyderman & Rothman 1988: 129).

Political scientist Amy Gutmann (currently president of the University of Pennsylvania and formerly provost of Princeton University) also claims that "the heritability theories of Jensen and Herrnstein have already been *quite thoroughly refuted*" (Gutmann 1980: 45 – italics added). For "forceful and cogent rebuttals" of Jensen's and Herrnstein's views on heritability, Gutmann refers the reader to Block and Dworkin, Kamin – and "especially" to three articles from the left-wing magazine *Dissent*. Her diagnosis reveals a woeful ignorance of the relevant literature. This should be obvious from many other things I say in this book, but let me just make two quick points here.

First, if Gutmann is right that Jensen's views should be completely rejected, how could she then explain the fact that Jensen's paper that started the whole controversy was, at the time of her writing, one of 100 most-cited articles in scholarly journals in social science (Garfield 1978: 652)? True, many of these citations were made for the purpose of disagreeing with Jensen, but in that context even a disagreement is usually a sign that the view is not devoid of all merit. In contrast to the practice in newspapers and political magazines, in peer-reviewed journals the utterly demolished theories tend to be ignored, rather than being condemned again and again. Therefore, we can be pretty sure that a "quite thoroughly refuted" theory is quite unlikely to become a citation classic.

Second, if Gutmann is right that Herrnstein's claim about a genetic component in IQ differences between socioeconomic groups is so conclusively disproved, how could she then explain that the very same "cogently rebutted" idea is also defended in, of all places, one of the mostly widely used textbooks in human genetics? Curt Stern, the then professor of genetics at the University of California (Berkeley), stated in 1973[16] that empirical data "suggest strongly that environment is not the sole agent – that there *are* mean differences in the genetic endowment of the different socioeconomic groups" (Stern 1973: 770 – italics in the original).

Another example of philosophers' bias is Ned Block's odd selectivity in his discussion of Flynn's views. As we saw, Block endorses Lewontin's "master-argument" without any reservation, but inexplicably, in this context he never mentions Flynn's detailed criticism of that same argument. When Flynn develops his own line against Jensen, however, his ideas suddenly become "interesting," and Block draws on them enthusiastically in his attack on hereditarianism. But he fails to make it known that Flynn himself would strongly disagree with the idea that hereditarianism should be dismissed because of its "conceptual confusions" and "flawed logic." On the contrary, Flynn regards Jensen as a formidable opponent whose work presents an extremely serious challenge to environmentalism. In

---

[16] Actually Stern had already made that claim in the first edition of his textbook (in 1949), and was criticized for it in the journal *Science* by a sociologist with the same last name (Stern 1950a). But Stern (the geneticist) remained unconvinced (Stern 1950b), and kept defending the same view in later editions.

the very essay which Block cites and uses as a *machine de guerre* against Jensen, here is what Flynn has to say about the debate between Jensen and his environmentalist adversaries:

> The result is something of a massacre, with Jensen showing that the most cherished environmental hypotheses have been sheer speculation without a single piece of coherent research in their favor. For this alone, all seekers of the truth are greatly in his debt. (Flynn 1987: 222)

With this kind of information omitted in Block's presentation, readers unacquainted with the literature will undoubtedly get a highly distorted picture of the controversy. It is as if Flynn's views have been passed through a filter that lets through only the ideas that can be used against Jensen and *The Bell Curve*.

The same bias is discernible in Block's short summary of his argument about within-group heritability and the racial gap in IQ (Block 1995: 115). In that summary Block mentions *only* three pieces of empirical evidence: the so-called Flynn Effect, the data about caste-like minorities, and the relatively small amount of genetic variation between the races. Notoriously, these are all the data standardly used to support environmentalism. Whereas these data are scrupulously discussed in *The Bell Curve* and then weighed against the *contrary* empirical information, more consistent with hereditarianism, in Block's concise picture of the controversy there is no place at all for the evidence that threatens environmentalism.

Block does briefly discuss some empirical data favoring hereditarianism but he seems to treat them as independent arguments (and not as an essential part of the inference from WGH to BGH). He faces the dilemma mentioned earlier. If he excludes the empirical component of the two-step argument in favor of hereditarianism, he is refuting the mere shadow of hereditarianism and is tilting at windmills. If he includes it, however, the charge of "conceptual confusions" and "flawed logic" simply disintegrates.

There is another manifestation of Block's partiality. Flynn's argument against Jensen (that Block endorses) is based on an observed increase of average IQ test scores in many countries in the last several decades – the phenomenon known as the "Flynn Effect." Flynn's argument about race builds on this and goes as follows. Since this secular gain in IQ must be due to some environmental causes (about which we are now completely in the dark), then there is also a hope that the racial gap in IQ might be accounted for in the same way (in terms of some presently unknown environmental differences). This is a possibility, but it is hardly to be taken

147

as immediately convincing or methodologically unproblematic, especially not in a contribution that aspires to examine the whole debate in a critical spirit. There are several obvious worries about the argument. For example, psychologist Robert C. Nichols raises a serious objection. He first reconstructs Flynn's argument and presents it in the form of four premises and the conclusion, and then dismisses it as a "faulty syllogism," or as an *obscurum per obscurius* reasoning.[17] But although this kind of criticism ought to be specially congenial to philosophers of science, and is moreover published as an immediate response to Flynn's paper on which Block so strongly relies, Block's paper contains no mention of that highly relevant contribution to the discussion, nor of any other of the conspicuous problems with Flynn's reasoning. Again, the filter blocks that kind of information.

4.5 THE HEREDITARIAN STRIKES BACK

As we saw, the main charge against hereditarians is that they fallaciously derive between-group heritability from within-group heritability. I tried to show that there is no evidence that hereditarians ever defended that inference. On the contrary, they very often explicitly condemn it because they are aware that WGH alone is insufficient to establish BGH.

Ironically, however, it is the anti-hereditarians who really have made a crude mistake of a similar kind. While the alleged error of hereditarians was that they supposedly inferred high between-group heritability from the strong genetic impact on IQ differences within groups, anti-hereditarians do sometimes argue *against* high between-group heritability of IQ on the grounds that the genetic variation is smaller between groups than within groups.

In discussing possible explanations of the black–white IQ difference Ned Block states that one of the four most important facts in this context is that "only about 7 percent of all human genetic variation lies between the major races"[18] (Block 1995: 112, 115). Block just lists this percentage

---

[17] "By a strange twist of logic Flynn has transformed the genuine mystery concerning test score changes over time into positive evidence that solves the alleged mystery of racial differences" (Nichols 1987: 234). Other scholars are also trying to stimulate "healthy skepticism" about the Flynn Effect (e.g., Rodgers 2000), warning that the phenomenon itself is at present so poorly understood that we should first strive to grasp better its nature and meaning, and only then attempt to explain it (or, for that matter, use it to explain something else).

[18] On p. 115 there is a typo: the words "within races" should obviously be "between races."

(taken from Lewontin) together with three other facts that in his opinion undermine the inference to the heritability of the black–white gap in IQ, but he does not explain at all how this particular division of *general* genetic variance (of a number of arbitrarily chosen loci) into components (within races and between races) becomes relevant for the *specific* question about IQ.

I think any argument of that form is logically flawed. For even if races differ genetically only to a small extent *on average* (taking into account a selected number of protein loci or genetic markers), this in itself is not a good reason to think that the races will also differ genetically only to a small extent *with respect to a specific phenotypic trait*. Block's attempt to use the low percentage of the between-races genetic variation ("*only* 7 percent) in his anti-hereditarian argument about IQ is fundamentally misconceived. If, averaged over many genetic loci, the racial variation constitutes only 7 percent of the total variation, it by no means follows that the proportion of racial variation *in IQ genes* will probably be 7 percent, or around 7 percent. It may well be, of course, but this cannot be established *a priori*.

Lewontin himself offers the following argument:

> Well, it might have turned out that there were big genetic differences between groups, and that most genes were highly differentiated between the major races. Now, if that turned out to be true, then at least it would be a possibility, although not demonstrated, that there might be, as some like to dream, high differentiations between groups in their mental abilities or in their temperaments or anything like that. Although nobody knew about any genes for those things, at least it was a living possibility. But when we found that there were practically no genetic differences between groups except skin color and body form and a few things like that, it became a great deal less likely and less interesting to talk about genetic differences between groups. (Lewontin 2003)

Lewontin's starting point is: (A) if it turned out that there were "big genetic differences between groups," then at least it would be a possibility that there might be genetic differences between groups in their mental abilities. Fair enough. But after the possibility of *big* genetic differences between groups is considered in the first step, we naturally expect to hear what would follow about between-group differences in mental abilities if the between-races genetic difference were found to be *small* ("only" 7 percent). Instead, Lewontin continues with the following claim: (B) "when we found that there were practically no genetic differences between groups except skin color and body form and a few things like that, it became a great deal less likely and less interesting to talk about genetic differences between groups."

Notice that no evidence whatsoever is offered to support the first part of (B), namely that practically the *only* genetic differences between groups are in skin color and body form. This is in fact the very issue that Lewontin was supposed to resolve by argument! The whole section is phrased as if (B) is argumentatively linked to (A), but this is obviously not the case. Moreover, although the main thrust of (A) is to prepare the ground for a refutation of the idea that there are genetic differences responsible for between-group differences in mental abilities, in the end Lewontin says nothing that would speak to that issue.

Since Lewontin mentions differences in skin color, a good question here is: do we expect that the component of inter-racial genetic variation with respect to *that* trait will be also around 7 percent? Certainly not. Actually, according to a recent study (Relethford 2002) it is 88 percent. Now the issue we are addressing is the following: is the distribution of genetic variance with respect to cognitive abilities more like (1) the case of skin color, where between-race variation is comparatively high, or like (2) genetic loci examined by Lewontin and others (Lewontin 1972; Barbujani et al. 1997), where the average between-group component is comparatively low (less than 12 percent), or perhaps (3) somewhere in between?[19] The honest answer is that we just don't know. This is an empirical question, and drawing inference about cognitive abilities on the basis of what we know about, say, blood groups is completely unjustified.[20]

Richard Dawkins and Steven Pinker fall into the same trap:

> the Human Genome Diversity Project (HGDP), now under way, builds on the foundation of the HGP but focuses on those relatively few nucleotide sites that *vary* from person to person, and from group to group. Incidentally, a surprisingly small proportion of that variance consists of between-race variance, a fact that has sadly failed to reassure spokesmen for various ethnic groups, especially in America (Dawkins 2003: 31).

---

[19] It is worth mentioning that even in Lewontin's own sample the variation of some genes showed a substantial between-race component. For example, the variance in *Duffy* and *Rh* was more than 25 percent between races (Lewontin 1972: 396; Lewontin 1982: 123).

[20] It is the same kind of mistake as if someone observed that two computers did not differ much, on average, on a number of arbitrarily chosen characteristics like size, color, weight, motherboard configuration, etc., and then concluded from this that the computers probably did not differ in the speed of their processors.

The quantitative differences are small in biological terms, and they are found to a far greater extent *among* the individual members of an ethnic group or race than *between* ethnic groups or races. These are reassuring findings. (Pinker 2002: 143)

If Dawkins and Pinker are right that Lewontin's figure is "reassuring," presumably because it points to the unimportance of genetic differences between groups, then by symmetry it should be possible in a relevantly similar situation to show the same thing about environmental causes. But I think we would all agree that in that case the reasoning would be fallacious.

Imagine that a hereditarian counterpart of Lewontin comes upon the scene, and that he undertakes careful measurement of many environmental influences within groups and between groups. Suppose that he eventually finds out that in his sample, average environmental variation in most traits he measured is much smaller between groups than within groups (say, between-groups variance is "only" 7 percent of the total variance). He then starts arguing that this is bad news for environmentalists and that "spokesmen for various ethnic groups" should be worried because the proportion of between-group environmental variance is "surprisingly small," and that "obviously" environmental causes cannot explain the between-group difference in cognitive abilities.

It is very clear what is wrong with this argument. The fact that between-group environmental differences do not have much impact *on average* does not show that they do not have much impact *on cognitive abilities*. In this context, we recognize the error immediately, but in the genetic case, although we are dealing with the same logical fallacy, we are more easily deceived.

Speaking about the quantitative comparison between within-race and between-race genetic variation, it should be pointed out that Lewontin used the results of his measurements to derive an unjustified conclusion not only about IQ differences between races but also about racial classification itself. He concluded that racial classification is of virtually no genetic or taxonomic significance, on the grounds that the genetic differences between the races on a number of arbitrarily selected loci were typically swamped by the corresponding within-race differences. But as A. W. F. Edwards recently showed (Edwards 2003), Lewontin completely ignored the correlations that exist among different loci, which actually support a racial taxonomy even in the absence of a very big average variation between the races on a gene-by-gene basis.

There are two puzzling facts here. First, it is odd that Lewontin never mentioned this obvious difficulty for his overhasty deconstruction of the concept of race, although he was undoubtedly aware of it.[21] Second, it is even more odd that during the last more than thirty years no geneticist has ever explained to the public this fatal weakness of Lewontin's widely disseminated and widely praised argument. Edwards, being apparently the first one to do this, said that the article in which he exposed Lewontin's statistical fallacy "could, and perhaps should, have been written soon after 1974" (Edwards 2003: 801). Indeed.

To sum up, here is the picture that emerges from this chapter. On one hand, Lewontin often claimed that hereditarianism is based on an elementary fallacy (inferring between-group heritability from within-group heritability). Although his accusation was in fact both false and unsupported by any evidence, it was nevertheless widely accepted by scholars and many others for more than thirty years. On the other hand, Lewontin himself did really make a simple inferential error (by concluding that if populations overlap on a number of characteristics, taken separately, they cannot be clearly distinguished on a combined measure). Although this reasoning is demonstrably wrong (and had been recognized as such by R. A. Fisher as early as in 1925), scientists have for decades hailed this fallacy as the most important "truth" about biology and race. It even made its way into the journal *Nature* (see the reference in Edwards 2003) and into "educational materials about genetics and genomics for students, teachers and the general public," which were issued by the National Human Genome Research Institute at the National Institutes of Health (http://www.genome.gov/Pages/Education/Kit/main.cfm?pageid=202).

How could all this happen? The short answer is that, evidently, some tremendously strong, mind-bending force must have been at work. The long answer? Well, for this we will have to wait for some brave soul to undertake this fascinating research project in the history of contemporary science.

---

[21] The point was made, among others, by Cavalli-Sforza and Edwards at a genetics congress in 1963 (see Edwards 2003: 799, 801). Lewontin himself also attended that congress but when he developed his argument against the race concept ten years later he failed to mention their analysis, which in fact undermines his reasoning.

5

Genes and malleability

Best-selling novels rarely have unhappy endings; similarly, books about genetics and social science usually close with some kind of sugarcoating about how biological traits are not really determined, or how a heritable trait is malleable.

David C. Rowe

It is not true that everyone can reach the same academic standards if provided with adequate opportunity, and the heritability of IQ is a partial measure of that untruth.

John Thoday

5.1 GENETIC AND ENVIRONMENTAL CAUSATION

Can phenotypic differences arising from genetic differences be eliminated as easily as environmentally caused differences? Those who answer this question in the affirmative like to point out that being caused by genes does not entail being unchangeable, fixed, or predestined. This trivial truth is easily granted. But after we concede that, indeed, "heritable" does not mean "unchangeable," there is a temptation to make another step from this truism to a much stronger claim, namely that *there is no difference at all between the ways genetic and environmental effects are modifiable.* This is a step from truth to falsity.

Let us begin with quotations from Jencks, Dawkins, and Lewontin, which make the same point and initially sound very plausible:

Most of us assume that it is harder to offset the effects of genetic disadvantages than environmental disadvantages. Because our genes are essentially immutable, we assume that their consequences are immutable too. Because

153

the environment is mutable, we assume that its effects are equally mutable. But there is no necessary relationship between the mutability of causes and the mutability of their effects. (Jencks 1988: 523)

Genetic causes and environmental causes are in principle no different from each other. Some influences of both types may be hard to reverse; others may be easy to reverse. Some may be usually hard to reverse but easy if the right agent is applied. The important point is that there is no general reason for expecting genetic influences to be any more irrevocable than environmental ones. (Dawkins 1982: 13)

Inherited disorders can be treated and corrected as easily (or with as much difficulty) as those arising from birth traumas, accidents or environmental insults. To cure a disease, we need to know what the metabolic or anatomical lesion is, not whether it is the result of being homozygous for a gene. (Lewontin 1976b: 8)

To plant a seed of doubt about this Jencks–Dawkins–Lewontin argument, let's take an example of an effect P that is genetic in the sense of being 100 percent heritable. This means that, under a given regime of parameters, the reason why some organisms have P is that they have a given genetic characteristic G_1, and the reason why other organisms lack P is that they have a different genetic characteristic G_2. If we want to modify trait P, then *under that regime* we can do nothing by environmental manipulation. There is, by definition, no role for the environment to make a difference. This shows that *in some sense* heritability does limit malleability.

But the point applies also to effects that are less than 100 percent heritable. The higher the heritability, the lower the degree of environmental modifiability. If an effect is, say, 60 percent heritable, then by exposing all organisms to the "best" of all existing environments it will be possible to eliminate at most 40 percent of variation in that effect.

So why did the argument to the contrary look so plausible? Well, I suspect that one reason was that it mixed two different things, the *emergence* of an effect and the *persistence* of that effect. If phenotypic characteristic P is genetic in the sense that its *emergence* is entirely explained by a genetic cause, obviously from this fact it does not follow that P will *persist* indefinitely, and especially it does not follow that there will be no way to change P by environmental manipulation. The initial assumption was that the emergence of P is genetic, not its persistence. Therefore, by making the distinction between two possible explananda we recover consistency: what is genetic (the emergence of P) is not readily modifiable, and what is readily modifiable (the persistence of P) is *ipso facto* not genetic.

To see that heritability puts constraints on malleability it is extremely important to keep constantly in mind which heritable (genetic) characteristic is exactly being considered. This is where Jencks, Dawkins, and Lewontin err. They all talk about diseases that are heritable or genetic, in the sense that their *onslaught* is caused by genes, and then they go on to stress that these diseases are nevertheless completely malleable, in the sense that their *continuous presence* can be stopped just by environmental manipulation. But these are two different things, and manifestly there is no intrinsic relationship between them. The whole point is that Jencks, Dawkins, and Lewontin should decide whether they want to talk about onslaught or about continuous presence. Once they make up their minds about this, they will see that, whichever of the two things they focus on, there will be a connection between heritability and limits on malleability. Simply, if the heritability of the onslaught of disease D is 100 percent, environmental manipulation within the existing range will not affect the onslaught of D at all. Likewise, if the heritability of the presence of disease D is 100 percent, environmental manipulation within the existing range will not affect the presence of D at all.

Take two examples, A and B.

(A) Consider heart problems that are due to congenital disorders but are curable through environmental intervention (surgery). Now if we ask to what extent we can decrease the number of people who need medical attention because of this kind of heart trouble, the fact that its cause happens to be fixed shows that, barring genetic manipulation, we cannot do much. This medical problem is genetic with respect to its origin but environmental with respect to its persistence.

(B) Imagine, in contrast, an environmentally caused disease that usually goes away on its own except in people with a given genetic constitution, in which case it is incurable. (I don't know whether such a disease really exists.) This disease would be environmental with respect to its origin but genetic with respect to its persistence.

Table 5.1 represents two questions, "Genetic?" and "Environmentally manipulable?", asked about two different things, onslaught and continuous presence. If these questions are answered separately for examples A

Table 5.1

	Genetic?	Environmentally manipulable?
Onslaught	Yes (A), No (B)	No (A), Yes (B)
Continuous presence	No (A), Yes (B)	Yes (A), No (B)

and B, we see that the incongruous combination of genetic *and* environmentally manipulable disappears.

It is now easy to locate the fallacy committed by Jencks, Dawkins, and Lewontin. They all start with an example like A. But then they take the answer to the question "Genetic?" from the first row, and the answer to the question "Environmentally manipulable?" from the second row, without realizing that these are answers to two completely different questions.

It is true that "environmental" does not automatically mean "modifiable," because there are some environmental causes that we don't know how to manipulate. But at least we know what would have to be changed in order to produce the desired effect. With genetic causality, we often don't know even that. It is not just that we don't know how to manipulate environments to change the effect; we don't even know which environmental manipulations would produce the desired effect (or whether such environmental manipulations exist or are discoverable at all).

Daniel Dennett considers the idea that the environment is a "more benign" source of determination because we can change the environment, but he immediately dismisses it on the following grounds:

> That is true, but [1] we can't change a person's *past* environment any more than we can change her parents, and [2] environmental adjustments in the future can be just as vigorously addressed to undoing prior genetic constraints as prior environmental constraints. And we are now on the verge of being able to adjust the genetic future almost as readily as the environmental future. (Dennett 2003a)

Concerning [1], yes, *past* environments are as unchangeable as genes received from the parents, but (currently at least) environmental sources of *future* phenotypic differences are much more under our control than genetic sources of these differences. Concerning [2], yes, it may well be that, as Dennett anticipates, "environmental adjustments in the future [will] be just as vigorously addressed to undoing prior genetic constraints as prior environmental constraints," but until this really happens, our different attitude toward genetic and environmental causation will remain justified, and particularly so in those numerous cases where, as of now, no environmental manipulation is available for "undoing prior genetic constraints."

It is ironic that despite subscribing to the complete symmetry of genetic and environmental causes, Dennett has been unable himself to stick to that symmetry consistently even within a couple of pages. Discussing Jared Diamond's book *Guns, Germs and Steel* (Diamond 1997), he says: "Is [Diamond] a dread genetic determinist, or a dread environmental determinist? He is neither, of course, for both these species of bogeyman are as mythical as werewolves." But just on the previous page, Dennett describes the content of that book in such a way as if he *did* regard the idea of a possible genetic difference between, say, European and African people as something of a bogeyman. Namely, he reassures the reader that Diamond did not entertain "some *awful* racist hypothesis about European genetic superiority" (Dennett 2003a; Dennett 2003b: 160 – italics added). But why should Dennett consider this hypothesis "awful"? Apparently, he forgot his own questions, raised only a few pages earlier, which suggested that there is nothing particularly dreadful about genetic explanations: "What would be so specially bad about genetic determinism? Wouldn't environmental determinism be just as dreadful?" Clearly, according to the very thesis of symmetry that he defends, he ought not regard a genetic account of Western dominance as any more "awful" than a purely environmentalist account.

Dawkins is similarly inconsistent. In accordance with the thesis of complete symmetry between genetic and environmental causes, he says that "there is nothing peculiar about *genetic* determinism which makes it particularly sinister" (Dawkins 1998 – italics added), yet (as mentioned in section 4.5) he finds it "reassuring" that average *genetic* differences between races are quite small (Dawkins 2003: 31). Well, if a small impact of genes is "reassuring," it follows that a large impact of genes would be upsetting. So, after all, it seems that even Dawkins himself agrees that there is something troubling about genetic influences, at least in some contexts.

5.2 PKU

It is of course possible that something that was in a given set of environments a purely genetic effect will cease to be so when novel environments are introduced, and that it will then become responsive to these new environmental influences. The case in point is the human disease phenylketonuria (PKU). People suffering from that genetic disorder are unable to digest phenylalanine (an amino acid that is a common ingredient in normal human diet), which then accumulates in their blood and usually leads to severe mental retardation. However, if they are put on a diet free of phenylalanine no disastrous effects occur. PKU is a favorite weapon in the fight against "genetic determinism." It is very often taken to show that heritability puts no limits on malleability, and that genetic effects are as easily modifiable as the environmental ones:

> PKU is one hundred percent hard-wired in the genes. Yet it can be effectively cured with a one hundred percent environmental intervention. (Collins et al. 2001)

> it is well known that heritability does not imply immalleability or fixity. A characteristic can possess a heritability of 100 percent and yet be fully remediable by environmental interventions. Perhaps the best known illustration of this point is phenylketonuria (PKU). (Lilienfeld & Waldman 2000)

> PKU is a genetically determined, recessive condition that arises due to a mutation in a single gene on chromosome 12 (with a heritability of 1), and yet its effects are highly modifiable. (Sternberg & Grigorenko 1999: 541)

> PKU is a trait with a heritability of 1.0. But its expression can be drastically altered by a change in environment. PKU thus demonstrates that biology is not destiny. (Paul 1998: 180)

> A related misconception is that the effects of a genetically determined abnormality cannot be changed by environmental manipulation. Again, this is wrong even with single gene diseases. The example of the inherited metabolic disorder phenylketonuria (PKU) illustrates this point well. (Rutter & Plomin 1997: 210)

> Heritability does not imply inevitability, because environment can determine the relative impact of genetic variation (GE interaction). For example, phenylketonuria – a genetic cause of mental retardation – is 100% heritable, yet affected individuals can avoid its consequences by eliminating phenylalanine from their diet. (Gray & Thompson 2004: 477)

(See also Jacquard 1984: 112; Alper 1998: 1604; Kurzban 2004.)

All the quoted authors defend basically the same position, and they certainly do not lack one thing: authority. Francis Collins is the leader of one of the two research groups that recently sequenced the human genome, and the others are also well-known protagonists in the nature–nurture debate. Yet they suggest something very implausible: that under the present conditions PKU is *both* 100 percent genetically determined *and* also strongly responsive to environmental influences. But to say that a variation is "genetically determined" or 100 percent heritable *means* that it is *entirely* due to genetic differences and that, under these circumstances, it is in no way affected by environmental variation.

Therefore, without even going into biological details, logic alone guarantees in advance that a trait with a heritability of 1 will *not* be environmentally modifiable, if we stay within the range of parameters for which heritability was calculated. Collins's claim that PKU is *now* "one hundred percent hard-wired in the genes" and that *at the same time* it "can be effectively cured with a one hundred percent environmental intervention" verges on self-contradiction. Also, when Gray and Thompson say that heritability can be 100 percent and yet "environment can determine the relative impact of genetic variation (GE interaction)," this does not make sense. Heritability of 100 percent is logically incompatible with the presence of GE interaction.

Let us look more closely at the case of PKU to explain how the mistaken view may have originated.

The full development of PKU consists of the following three steps in a causal sequence: (1) a gene mutation → (2) inability to metabolize phenylalanine → (3) mental retardation. Now, if someone says that PKU is both 100 percent heritable and environmentally modifiable it is crucially important to know which effect he has in mind: (2) or (3). Let's explore both possibilities (see Figures 5.1a and 5.1b). On one hand, (2) is indeed 100 percent heritable because the difference between those who are unable to metabolize phenylalanine and others is entirely explained

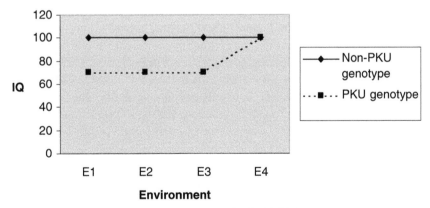

Figure 5.1a Is PKU 100 percent heritable? Not any more.

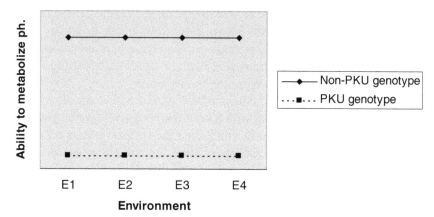

Figure 5.1b Is PKU 100 percent heritable? Yes, it still is.

by a genetic difference. But, under the present conditions this defect cannot be removed by environmental manipulation. (There is no treatment that can help them metabolize phenylalanine.) On the other hand, (3) is modifiable environmentally because mental retardation can be avoided by an appropriate diet (free of phenylalanine). But by this very fact, (3) is not 100 percent heritable. The difference between those having this particular kind of mental retardation and others is explained by a genetic-cum-environmental difference. Strictly speaking, the trait is no more heritable than environmental.[1] It is due to the statistical interaction of genes and environments, in the technical sense of that term (see chapter 2 for extensive discussion).

So, in the end everything is as it should be, in perfect logical order: the oxymoron about totally heritable yet readily modifiable traits disappears. Briefly, PKU involves two effects: a metabolic defect and a psychological disorder. The former is 100 percent heritable and (currently) non-modifiable, whereas the latter is highly modifiable but its heritability is low. Clearly, if the idea was to argue that wholly heritable traits can be environmentally plastic on the grounds that, in the context of PKU, one thing (the metabolic defect) is wholly heritable and the other thing (the psychological disorder) is still avoidable through appropriate environmental measures, this is a serious confusion and equivocation. Referring

---

[1] Cf.: "Although PKU is a 'genetic condition' [notice the scare quotes], mental retardation among people with PKU manifests only as a result of the interaction of genetics and environmental exposure (in this case, diet)" (Shostak 2003: 2331). Also: "PKU retardation is an example of a Genetics × Environment interaction" although it "originally *appeared* empirically to be a genetic main effect" (McCall 1991: 146 – italics added).

to the three-link causal sequence, if in speaking about PKU you just make sure that you are always consistent and that you know exactly whether you want to talk about (2) or (3), you will realize that nothing remains that is both 100 percent heritable and "fully remediable by environmental interventions."

There is, however, another source of the ill-conceived attempt to use PKU to show that heritability puts no constraints on malleability. Now the idea is to focus just on the PKU-related mental retardation, and argue that it became environmentally modifiable despite previously having been maximally heritable (i.e., genetically caused). This is indeed true, but in a rather trivial sense.

To see this, it is best to look at how our knowledge of PKU advanced through time. At the first stage, the only thing known was that the drastically low IQ in this case was a heritable disorder (due to a recessive gene), and there was no way to forestall the mental retardation that was a typical outcome for homozygotes. Heritability was 1, and environmental modifiability zero. Later, at the second stage, the key role of the accumulation of phenylalanine was recognized, and mental retardation became avoidable. But with this change, as already pointed out above, heritability was no longer 100 percent. Namely, at present the difference between those who have that specific kind of mental retardation and those who do not is no longer explainable purely by a genetic difference. Rather, it is now regarded as the *joint* effect of heredity (gene) and environment (diet). In statistical terms, the variation is largely due to gene–environment interaction.

Why did heritability change? Well, because heritability is a population-relative statistic, and among other things it depends on the range of environmental variation. The reason heritability was 100 percent at the first stage is that at that time no environmental influence made a difference with respect to the presence/absence of the mental retardation, which was a common fate for people with two recessive genes for PKU. At the second stage, heritability dropped significantly because environmental variation was extended in a highly significant way: a new and critically important environmental influence (the phenylalanine-free diet) was discovered that broke the previously uniform connection between the PKU genotype and mental retardation. Now the effect of a person's PKU genotype became contingent on the kind of environment he had been exposed to.

The message is again that, as should have been obvious in the first place, the incoherent combination of full heritability and easy modifiability cannot arise. This time everything falls into place once we make a distinction between the two phases in our understanding of PKU. Initially, heritability was 1, but modifiability was 0. Later, the trait became modifiable but heritability decreased as well. So, no mixing of contradictory opposites here, no squaring of the circle.

It should be pointed out that although PKU is a favorite weapon against hereditarianism, it has been used to criticize heritability in ways that are logically incompatible. For the authors quoted above, PKU is supposed to be an instance of a trait that is *both* entirely heritable but currently malleable, but on other occasions the very same disease is presented as showing that in such cases assigning heritability simply makes no sense (e.g., Medawar 1977a). Necessarily, at least one of these claims must be false.

I tried to show that the case of PKU is misrepresented when it is used to convey a "feel-good" and optimistic message that genetic effects are readily alterable. Interestingly, it has been argued that the simple success story usually told about PKU does not correspond to the much more complicated empirical reality of that disease (Hay 1985: 61–63; Paul 1998). Diane Paul's diagnosis of why the oversimplified description has been so popular has a ring of truth: "Both enthusiasts for genetic medicine and critics of genetic determinism have come to find the story useful. These convergent interests mean that no one has an incentive to pick up the rock and see what lies underneath" (Paul 1998: 180).

5.3 LOCAL MODIFIABILITY AND MODIFIABILITY "IN PRINCIPLE"

Here is a clear example of how the link between heritability and limited malleability is attacked by confusedly running together heritabilities at *different* times: "The heritability of a trait gives us no hint as to how malleable or plastic the trait may be. Before the discovery of the faulty copper metabolism that underlies Wilson's disease, it *was* 100 percent heritable" (Lewontin 1976b: 10 – italics supplied). Yes, indeed, it *was* heritable then, but it *is* no longer heritable. No one would dispute the trivial truth that the magnitude of heritability can vary over time due to changes in the relevant population parameters, but it still remains indisputable that, within a given regime, heritability is incompatible with unlimited malleability.

Lewontin makes a similar statement in another article, again trying to show that it is the "chief programmatic fallacy" to think that the heritability of some trait is an index of environmental modifiability: "A trait can have a heritability of 1.0 in a population *at some time*, yet could be completely altered *in the future* by a simple environmental change" (Lewontin 1976a: 179 – italics added). The italicised phrases disclose that although Lewontin thought that, with this statement, he addressed the issue of heritability and malleability, what he actually said was merely that heritability in a given population could be different at different times. Who would disagree with this platitude?

What about heritabilities lower than 100 percent? The basic point still applies: the more heritable a variation, the less environmentally modifiable it is. Contrary to what critics of hereditarianism so often say, heritability does place a constraint on malleability (if environmental influences are restricted to those already included in the calculation of heritability). For example, if heritability is 60 percent and environmentality is 40 percent, then by equalizing existing environments you can eliminate only 40 percent of the total variation. If you aim for more with this method, you are logically confused.

All this does not mean, of course, that heritable differences are written in stone, and that they are *absolutely* unmodifiable. It is always possible that something that was a purely genetic effect in a given set of environments will cease to be so when novel environments are introduced, and that it will then become responsive to these new environmental influences. But this possibility does not prove, as is frequently implied, that genetic effects are as easily modifiable as the environmental ones. The crucial structural asymmetry between the two may be described as follows. Environmentally produced phenotypic variation can be leveled out by exposing all organisms to the same kind of environment, *out of those already known*. On the other hand, barring the policies that put environmental handicaps on those with a genetic advantage, genetically caused (heritable) phenotypic differences can be removed only if a new environmental difference is found *outside the existing environmental range* which will interact with genetic influence and which could be used to offset it. Needless to say, in any particular case one simply cannot be certain that such an environment exists, or if it does that it is just about to be discovered.

Discussions about heritability and malleability should be governed by the same rules as other discussions about modifiability. What we are usually interested in is the possible impact of those measures that are at our disposal or those measures that we can reasonably anticipate will be developed in the near future. We do not usually call something modifiable if there is only a logical possibility that it will come under our control. For example, when we say that Alzheimer's disease is incurable, we mean, roughly, that no current medical intervention can stop the degenerative process in the brain that leads to death in about seven to ten years. What strengthens the claim of incurability is that there have been intensive attempts to find a cure, which have produced no results so far. So, the existing environmental variation includes the measures that were deliberately introduced in the hope that they might be effective. It is precisely the fact that these measures were unsuccessful that justifies the modal claim that the disease is incur*able* (i.e., that it *cannot* be cured, rather than that it is just not cured).

We would be completely baffled if someone criticized the statement that Alzheimer's disease is incurable by saying that certain effective, though presently unknown, interventions might become available some day. Of course they might! Who would deny that? Surely, the word "incurable" does not mean "something that has no cure now, and is bound to remain without cure in all eternity and in all possible worlds." If, *per absurdum*, it did mean that, the word would be totally useless.

But those who object to the claim that heritability constrains malleability often commit the same mistake of replacing the context-relative and only meaningful way of discussing malleability or modifiability with a trivially true but irrelevant statement that heritable traits are not *absolutely unchangeable*. Here is an example how the debate is muddied by introducing an unnecessarily strong word, "immutability":

> There continues to be a popular but mistaken belief that the level of heritability equates with the ease or difficulty of changing or altering a particular characteristic, or its *immutability*. However, researchers in behavioral genetics and psychologists would now agree that the ways in which different factors interrelate in the development of a characteristic are not related to its *immutability*. (Nuffield 2002: 21 – italics added)

Notice how the sensible question whether the level of heritability is related to "the ease or difficulty of changing a particular characteristic" (to which one might well be disposed to give the answer "yes") quickly shifts to, and is indeed replaced by, a pseudo-issue of "immutability." Difficult to change? Perhaps yes. Currently unmodifiable? Again, maybe yes. But immutable? Of course not!

Let us call "locally modifiable" those phenotypic differences that can be eliminated by manipulating environmental influences in the existing range, and "in principle modifiable" those differences that can be eliminated only by finding a new environmental influence, outside the existing range. Then, environmentally caused phenotypic differences are locally modifiable, and heritable differences are not. Although heritable differences are "in principle modifiable," it is easy to see that the much more interesting sense of "modifiable" is "locally modifiable."

Here is a good explanation of social implications of high heritability and of its impact on local modifiability:

> the great majority of immediate policy decisions revolve around just that set of environments for which heritability estimates have the most relevance: the existing set. Most proposed policy changes involve minor redistributions of environments within the existing range, and it is precisely regarding such changes that a heritability estimate has its maximum predictive value. For instance, one message that a high heritability coefficient can convey is that minor fiddling around with environmental factors that already vary widely within the population has poor odds of paying off in phenotypic change – and thus new ideas about environments need to be tried. Surely, this is a message of enough social and practical implication to justify continued interest in heritability and its estimation. (Loehlin et al. 1975: 99)

But as explained above, the expression "local modifiability" gains on strength and relevance if "existing environments" involve many attempts to influence the trait in question. If these attempts are unsuccessful and if the heritability remains high, then we will know not only that redistribution of existing environments will have little effect on phenotypic differences but also that all the concerted efforts undertaken so far to influence the phenotype have failed. In that case, "locally non-modifiable" would move closer to what common sense understands as "non-modifiable."

Note, however, that even in that situation the trait would still be "modifiable in principle." But the trouble with the expression "modifiable in principle" is that it is devoid of any useful information. *Every* difference is in principle modifiable. For all we know, any difference could be eliminated by finding some yet unidentified new environmental influence. Who would be so foolish as to say about any kind of phenotypic difference that it is not modifiable in principle? No one has been. Those who insist that even heritable traits are in principle modifiable are wasting their time by arguing for an utterly trivial claim.

It is easy to agree that heritability does not imply immutability (Neisser et al. 1996: 86; Alper 1998: 1604; Ridley 2000: 85; Brown 2001: 193; Nuffield 2002: 21), mainly because "immutability" suggests something very implausible, an absolute impossibility of change. But all this should by no means be taken as justifying the following radical claims:

> *The heritability of such scores tells us nothing about the educability of the children being tested.* (Layzer 1972a: 270)
>
> Heritability has little or nothing to do with the malleability of the trait in question. (Blackburn 2002)
>
> There is no connection whatsoever between the variation that can be ascribed to genetic differences as opposed to environmental differences and whether a change in environment will affect performance and by how much. (Lewontin 1993: 29)

The claim is also endorsed in the DST manifesto: Oyama et al. 2001: 3.

There *is* a connection between heritability and modifiability, not in the sense that heritability excludes modifiability but in the sense that it puts significant constraints on it. Unfortunately this real and important implication of heritability tends to be obscured by the critics of hereditarianism who deflect the discussion from the real issue by needlessly fulminating against the idea (believed by nobody) that "IQ is inscribed in stone in the DNA of each one of us" (Pigliucci 2001: 255), that heritability entails inevitability (Gould 1981: 156), etc.

William R. Havender notices a similar fallacious move in Feldman and Lewontin's criticism of Jensen (Feldman & Lewontin 1975). They first present Jensen as claiming that high heritability shows the futility of *any conceivable* environmental interventions, when in fact he had in mind just environmental interventions of a *certain kind*. Then this distorted statement becomes "an easy, even unworthy beast to slay" (Havender 1976: 608).

In his attempt to slay the "vampire" of genetic determinism, Paul Griffiths also vacillates between attacking non-modifiability in principle and local non-modifiability. In two versions of his paper "The Fearless Vampire Conservator" he gives two different definitions of genetic determinism. In one version, he proposes the following strong interpretation: "Genetic determinism is the idea that many significant human characteristics are rendered *inevitable* by the presence of certain genes; that it is *futile* to attempt to modify criminal behavior or obesity or alcoholism by any means other than genetic manipulation" (Griffiths 2002b – italics added). This kind of determinism involving "genetic inevitability" is a straw man. In the second version he proposed a much weaker definition:

"Genetic determinism is the idea that significant human characteristics are *strongly linked* to the presence of certain genes; that it is *extremely difficult* to attempt to modify criminal behavior or obesity or alcoholism by any means other than genetic manipulation" (Griffiths 2002a – italics added). But now, so defined, the view becomes a contingent empirical claim. It may be true or false, depending on the trait in question, but it surely cannot be ruled out of court by a methodological "stake-in-the-heart" move.

Griffiths's talk about "genetic determinism" is not helpful at all. For although in the first case the label is justified (because of the words "inevitable" and "futile"), it describes a standpoint that no one has advocated. In the second case, though, the term is inappropriate precisely because the rigid connotation of "determinism" does not correspond to the view, which (as he describes it) is no longer so simplistic and inflexible. Given this mismatch between definiendum and definiens, can it be that Griffiths has here been hoisted by his own terminological petard? Can it be that by taking too seriously a mere label that already sounds too crude to be true, it is Griffiths himself who has actually conjured up the "monster" of genetic determinism, and that the vampire he is fighting exists only in his imagination?

There is another important issue that remains to be clarified. We defined as "locally modifiable" those phenotypic differences that can be eliminated by manipulating environmental influences in the *existing* range of environmental variation. But does local modifiability, so understood, correspond to what we usually mean by trait modifiability? In other words, how important is what happens within the existing range for our judgment about the modifiability of a given phenotypic trait? After all, what merely exists is not a good measure of what is possible. That is true. Nevertheless, the more the existing range of environmental variation includes intensive, systematic, and diverse efforts to change the trait in question, the more it comes to reflect the commonsense notion of modifiability. It is precisely for this reason that with traits like IQ, which have a long history of attempted environmental modifications, the high heritability gives important information about limits of their malleability.

If heritability were indeed irrelevant for malleability it would be difficult to explain why politically sensitive critics of heritability are then so eager to dispute the hereditarian claim. For, if they really believed their own "irrelevance thesis," they would be happy to just hammer in the point that, *whatever the value of heritability*, no conclusion follows about malleability, and they would consequently be entirely indifferent as to what the empirical estimate of heritability turns out to be. But they are not indifferent at all. With respect to some traits, they regard the genetic explanation of group differences as "awful" (Dennett), and they react with a sigh of relief if the evidence shows that the influence of genes is small, calling this discovery "reassuring" (Dawkins and Pinker). This reaction is inconsistent with their proclaimed view that it doesn't matter whether an effect is genetic or environmental.

Notice, however, that the attitude toward heritability shifts dramatically when, instead of an urge to *change* the status quo, there is an interest in *preserving* the situation as it is. In cases where people prefer to regard a given trait as "fixed" and not readily modifiable, high heritability becomes good news, exactly as one would expect if it is true that heritability constrains malleability. For example, many homosexuals welcomed the hypothesis about the genetic basis of homosexuality, precisely because they hoped that acceptance of that hypothesis would stop efforts to change their sexual orientation. They hoped that the genetic hypothesis would make homosexuality look like an *inherent* and "natural" psychological disposition, rather than as an "unfortunate" result of "bad" environmental influences, which could be manipulated with the purpose of eliminating the trait.

Careful readers will have noticed that in discussing malleability I was talking about phenotypic *differences*. The claim was that high heritability of a trait in a given population puts constraints on the malleability of genetically caused *differences* in that trait. The situation changes, though, when the issue is about the malleability of the phenotypic *mean* of that trait. Even a very high heritability, as long as it is not 100 percent, leaves open the possibility that manipulating the environmental differences in the existing range can lead to dramatic changes in the phenotypic mean.

Here is an illustration how this can happen. Imagine a population of a million people, half with genotype G_1 and the other half with genotype G_2. Let us also assume that, if the environment is fixed, G_1 leads to a phenotypic value two points higher than G_2. Furthermore, of the two environments (E_1 and E_2), E_1 leads to a phenotypic value 10 points higher than E_2 (if the genotype is fixed). Finally, imagine that only one person is exposed to environment E_1, while all others are exposed to E_2. Table 5.2 represents the phenotypic values of people with the four possible G–E combinations, and (in the parentheses) the number of people with each of these combinations.

With almost all the variation being in the right-hand side of the table (environment E_2), the phenotypic standard deviation will be very close

Table 5.2

	E_1	E_2
G_1	111 (0)	101 (500,000)
G_2	109 (1)	99 (499,999)

to 1. Put differently, virtually all organisms will have a 1-point deviation from the phenotypic mean ($\sim$100). Now the interesting thing is that, despite the environmental variation being negligible (and heritability being very high), environmental manipulation can nevertheless work miracles here: exposing all organisms to E_1 would increase the phenotypic mean 10 points, or 10 standard deviations.

Agreed. But we should not use this kind of situation to argue, as some people do, that high heritability puts no limits on phenotypic change.[2] Two things should be borne in mind. First, however impressive the result may be in changing the phenotypic *mean*, it still remains true that exposing all organisms to E_1 would have only a negligible effect on the phenotypic *variance*. The standard deviation of 1 would stay unaltered. High heritability does put a limit on the malleability of *differences*.

Second, if an environmental variable varies widely in a population and yet contributes little to phenotypic variation, then, other things being equal, manipulating this environment will neither have much impact on phenotypic differences *nor on the phenotypic mean*.

5.4 COMPARING APPLES AND ORANGES

David Layzer proposes the following argument that would lead to the surprising conclusion that heritable variation can be largely caused by environmental differences:

> if the environmental variable E is distributed in a narrow range about the value e_1 [as illustrated in Figure 5.2], h^2 is close to unity. Yet in these circumstances the phenotypic variance could reasonably be considered to be largely environmental in origin since it is much greater than the phenotypic variance that would be measured in an environment ($E = e_2$) that permitted maximum development of the trait, consistent with genetic endowment. (Layzer 1974: 1260)

---

[2] Lancelot Hogben used a structurally similar example to break the connection between heritability and limited malleability (Hogben 1933: 116–117).

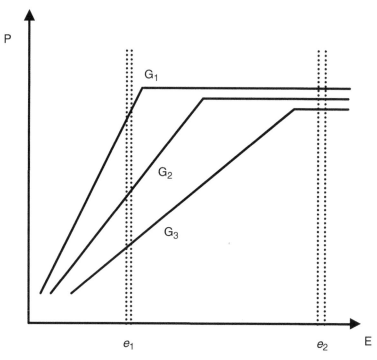

Figure 5.2 Confusion about range.

Layzer's mistake here is that he is running together two concepts that should be kept strictly apart (see Figure 5.2): (1) the phenotypic variation *within* either of the two narrow ranges (e_1 or e_2), and (2) the phenotypic difference *between* organisms exposed to e_1 and those living in e_2. Of these two variances, (1) is highly heritable, but (2) is not. For example, the phenotypic variation within range e_1 is indeed highly heritable because it is largely the result of the differences between the three genotypes (G_1, G_2, and G_3), while the environmental influence is in this context fairly weak. It can only be a confusion to argue, as Layzer does, that the phenotypic variation within range e_1 could be attributed to environmental differences on the grounds that "it is much greater than the phenotypic variance that would be measured in environmental range e_2." It is a matter of very simple logic that if your aim is to estimate the heritability of a phenotypic variation within a given range of environmental influences, you are obviously not allowed to consider anything outside that range. Therefore, any quantitative comparison *between* phenotypic variances in the two different environments, e_1 and e_2, must be utterly irrelevant for estimating the heritability *within* e_1.

The same fallacious reasoning is defended by Lerner (1986: 126) and Sarkar (1998: 84), who both reproduce Layzer's figure. Sarkar says:

> the phenotypic variability in [the range around e_1] should be attributed to environmental factors given that, for the same amount of genetic variability, there is so much more phenotypic variability at e_1 than in a range of environments (of the same size) around e_1. (Sarkar 1998: 84)

Notice that, like Layzer, Sarkar also assesses the size of environmental variance *within* e_1 by helping himself to an illicit look at a distant region around e_2. Both he and Layzer actually acknowledge that they are heavily indebted to Lewontin (1976c) for this criticism of heritability. Interestingly, it was in 1970 that Donald Hebb independently raised the same muddled objection:

> Mark Twain once proposed that boys should be raised in barrels to the age of 12 and fed through the bung-hole. Suppose we have 100 boys reared this way, with a practically identical environment. Jensen agrees that environment has *some* importance (20% worth?), so we must expect that the boys on emerging from the barrels will have a mean IQ well below 100. However, the variance attributable to the environment is practically zero, so on the "analysis of variance" argument, the environment is not a factor in the low level of IQ, which is nonsense. (Hebb 1970: 568; Cf. Hebb 1980: 77)

If there is any nonsense here, it is created by Hebb's not making the necessary distinctions. Again, two very different questions can be asked about Twain's tongue-in-cheek proposal. First, is the environment a factor in explaining the IQ differences between the boys raised in barrels? Clearly, the answer must be no, because the assumption is that their environment is "practically identical." Second, is the environment a factor in explaining the IQ difference between the boys raised in the barrels and those brought up normally? Now the answer is a resonant "yes." No contradiction, no paradox. (Hebb's argument is approvingly quoted in Lerner 1992: 112 and Lerner 1986: 128, with no mention of Jensen's reply [1971] in which Hebb's error is very clearly explained.)

The same mixing of apples and oranges occurs in a quite unexpected place, in one of the leading and most frequently used textbooks in genetics. Let me first reproduce a table from that book with "a hypothetical data set" that is supposed to illustrate the joint presence of 100 percent heritability and high environmental modifiability (Griffiths et al. 2000: 757).

Table 5.3

	Children	Biological parents	Adoptive parents
	110	90	118
	112	92	114
	114	94	110
	116	96	120
	118	98	112
	120	100	116
Mean	115	95	115

In Table 5.3, the numbers represent hypothetical IQ scores of a group of children, of their biological parents, and of their adoptive parents. (The parents' scores are the mid-parent values.) It is easily observed that the IQ differences between children are fully correlated with the IQ differences between their biological parents. The textbook authors (among whom is Richard Lewontin) conclude that the heritability of IQ among children is 1 ("heritability" being defined in a standard way as "the part of the total variance that is due to genetic variance"). The variation of IQ among children is completely explained by their genetic differences (inherited from their biological parents). Fine. But then Griffiths et al. add:

> Second, however, we notice that each of the IQ scores of the children is 20 points higher than the IQ scores of their respective biological parents and that the mean IQ of the children is equal to the mean IQ of the adoptive parents. Thus, adoption has raised the average IQ of the children 20 points higher than the average IQ of their biological parents; so, as a *group*, the children resemble their adoptive parents. So we have perfect heritability, yet high environmental plasticity. (Griffiths et al. 2000: 757)

The last sentence is logically incoherent. "Perfect heritability" means that phenotypic differences are *entirely* the result of genetic differences. "High environmental plasticity" on the other hand means that phenotypic differences are *not entirely* the result of genetic differences (i.e., that they are to a great extent the result of environmental differences). Therefore, although in the situation as described both parts of the sentence do initially sound plausible, the principle of non-contradiction tells us that they can in fact be simultaneously true only if they do not refer to the same range of phenotypic differences. And indeed, a quick analysis confirms that they do not.

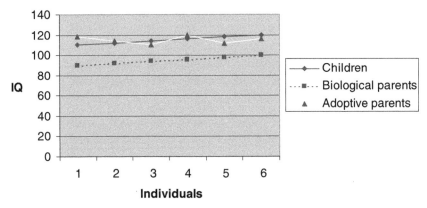

Figure 5.3 Whose heritability, whose plasticity?

Heritability of 1 refers to the IQ differences *within* the children group, but the environmental impact of adoption refers to the higher IQ of the children *as a group*. In the first case (perfect heritability), what we are looking at is the differences between the children (inside the first column). In the second case (high environmental modifiability), we are looking at the difference between the children (the first column) and their biological parents (the second column) – or alternatively, between the adopted children (the first column) and those children who are not adopted and who continue to live with their low-IQ biological parents (not represented in the table). As soon as we separate apples and oranges, consistency returns.

Looking at Figure 5.3, perfect heritability reigns in the differences between diamonds (children), but the difference between diamonds and squares (children and their biological parents) is completely environmental.

This whole conceptual confusion carries Lewontin's signature because basically the same example is similarly misused in Rose et al. 1984: 98–99, 116–117 (cf. also Lewontin 1993: 34). And, as with several other of Lewontin's arguments about heritability, this fallacy was also endorsed in the philosophy of science literature (Ariew 1996: S22).

Douglas Wahlsten, one of the most outspoken critics of heritability, is similarly adamant that malleability is not constrained by heritability:

> It does not matter whether the field of human behavior genetics finally decides that the heritability of IQ in the United States is 25%, 40%, 50%, or 70%. Any such estimate will be *utterly useless* to anyone seeking better ways to improve the intelligence of the nation through health care and education. (Wahlsten 1997: 84 – italics added)

Wahlsten combines this methodological point with an apocalyptic political prediction: the only use for heritability estimates that he envisages in the human context is "eugenic selective breeding," which is associated with "dismantling of democracy" and "descent into fascism." Salvador Luria also found a link between an interest in heritability and Nazism:

> Even if IQ were inheritable and its differences between races statistically significant, there is nothing sensible one can do about it, except possibly abolish the IQ tests (which may not be a bad idea) or improve school curricula (if one knew how) – unless, of course, what IQ enthusiasts want is to segregate the races: in schools, perhaps, or in concentration camps. (Luria 1999: 94–95)

But surely there are some less sinister implications of heritability. After all, it is a mathematical consequence of how heritability is conceptualized that it is inconsistent with the omnipotence of the environment. Non-zero heritability puts a precisely defined limit on what certain environmental influences can do. For instance, 70 percent heritability of IQ entails that equalizing environments in the range for which heritability was calculated *cannot* eliminate more than 30 percent of IQ variation in that population. Let me illustrate more concretely how the magnitude of heritability does matter by showing how some important parameters change when heritability takes different values: 0, 0.25, and 0.7.

In Table 5.4, the three rows show the variance of IQ, its standard deviation, and the range of IQ within which 95 percent of the population lies. You see how differently these three variables would change from the values they have in the actual world (column 4) if all existing environmental differences were eliminated – depending on whether heritability is zero (column 1), 0.25 (column 2), or 0.7 (column 3). (The two non-zero values are picked out because they are mentioned by Wahlsten.) It transpires that the effect of the radical environmental redistribution is quite modest in the case of high heritability (0.7), and also significantly weaker than in the case of low heritability (0.25).

Table 5.4

	E equalized $h^2 = 0$	E equalized $h^2 = 0.25$	E equalized $h^2 = 0.7$	Actual world
Variance	0	56.25	157.5	225
Standard deviation	0	7.5	12.5	15
95% within range	N/A	85–115	75–125	70–130

All this applies, of course, only to *existing* environmental variation. Heritability puts no constraints on untried, completely new environmental interventions. Hardly anything does. "Modifiability in principle" knows no limits, but this is precisely why, as an idea, it is vacuous and not worth discussing. It *is* worth discussing, of course, whether (and how) the differences that are at present locally non-modifiable could be made modifiable in the future, but it is unclear how reciting the trivial truth that they are "in principle modifiable" could help in this effort.

Put differently, the fact that heritability entails local non-modifiability surely does *not* mean that heritability somehow excludes or discourages attempts to eliminate heritable differences in the future. However, it does mean that as long (and insofar) as they remain heritable, they should be regarded as currently non-modifiable.

It is especially unhelpful when *merely possible* changes are used to dispute the claim that heritability constrains local modifiability:

> H & M [Herrnstein and Murray in *The Bell Curve*] claim high "heritability" of IQ means that improving the environment of a poor child a modest amount will be ineffective because "such changes are limited in their potential consequences when heritability so constrains the limits of environmental effects." They are simply wrong on this point. They commit what Lewontin has termed the "vulgar error that confuses heritability and fixity." A heritability estimate does not in any way "constrain" the effects of a changed environment. Bad genes and poor upbringing most certainly can impair mental development, but there is nothing about either genetic or environmental effects on the mind that renders them impervious to advances in scientific knowledge. (Wahlsten 1997: 73)

Wahlsten purports to prove that (1) "a heritability estimate does not in any way 'constrain' the effects of a changed environment," and his central reason supporting that conclusion is that (2) "there is nothing about either genetic or environmental effects on the mind that renders them impervious to advances in scientific knowledge." Well, of course, (2) is obviously true, but it amounts to "modifiability in principle," and it does not connect with the real debate. It is hard to think of a single effect that is known to be totally impervious to advances in scientific knowledge. For example, we surely cannot exclude the possibility that in the future a treatment will be discovered that would make people with Down's syndrome have an average IQ. It is unclear, though, how mentioning this mere possibility could in any way be illuminating (or comforting).

175

Take the following case of a baby with a serious genetic disorder, as described by its parents:

> In December 1986 our newly-born daughter was diagnosed to be suffering from a genetically caused disease called Dystrophic Epidermolysis Bullosa (EB). This is a disease in which the skin of the sufferer is lacking in certain essential fibers. As a result, any contact with her skin caused large blisters to form, which subsequently burst leaving raw open skin that healed only slowly and left terrible scarring. As EB is a *genetically caused* disease it is *incurable* and the form that our daughter suffered from usually causes death within the first six months of life . . . Our daughter died after a painful and short life at the age of only 12 weeks. (quoted in Glover 2001: 431 – italics added)

Would it be all right here to tell the child's parents: "Look, you are conceptually confused. It is wrong to think that because your daughter's disease is genetically caused it is also incurable. In fact, since at present only a small part of your daughter's genotype's norm of reaction is known, such fatalistic conclusions about ineffectiveness of environment are completely unjustified."

The reason this response sounds bizarre (and like a cruel joke) is that common sense is firmly on the side of the parents. The word "incurable" *never* means "incurable in principle," simply because in that interpretation it is a vacuous concept. Rather, it usually means "incurable now, with existing medical knowledge and despite repeated attempts to find a cure," i.e., when it matters to the speaker.

Since in this particular case the bare conceivability of an effective treatment would certainly not make us regard the genetic defect in question as curable or "environmentally malleable," why should we then allow ourselves to be pushed to draw the very same inference about other, relevantly similar, heritable differences? The word "pushed" is deliberately used here. Namely, in the heredity–environment controversy one is under distinct pressure to come up with a message that biological differences represent no barrier whatsoever to changes by environmental manipulation. If you defend, say, the heritability of IQ without at the same time taking pains to reassure the public that heritable does not mean unmodifiable, this is often enough to get you associated with genetic determinism, irresponsible pseudoscience, and status quo-ism in politics. Not that repeating the mantra "Heritability does not entail non-malleability" can guarantee that you will avoid these charges, but it will probably be perceived as a gesture in the right direction.

5.5 A CLUMSY ATTEMPT TO APPEASE THE CRITICS

This strong and generally felt pressure to de-emphasize some implications of heritability is the best explanation why certain authors who are at the very core of research in behavior genetics tend nevertheless to make dubiously consistent claims when addressing this topic.

For instance, Robert Plomin has been criticized by Susan Oyama for the following incoherence:

> Plomin states that genetic influence is not deterministic and "puts no constraints [*sic*] on what could be." Then he approvingly cites some colleagues' complaint that environmentalism has fed unrealistic expectations about malleability . . . It is hard to have it both ways, to argue that "genetic influence" places no constraints and that it precludes some outcomes. Before making sweeping statements about constraints, people should be clear about what they mean. (Oyama 1988a: 98)

Oyama thinks that Plomin's apparent inconsistency is the outcome of a half-baked and poorly thought out theoretical position. I would suggest an alternative explanation, mainly because it seems to me that the issues are here too simple for an authority of Plomin's caliber to be conceptually confused about. What must have happened, in my opinion, is that Plomin was just torn between two conflicting goals. On one hand, he wanted to acknowledge the important point that substantive heritability is indeed incompatible with unlimited malleability, and hence he had to state that heritability does decrease local modifiability. On the other hand, I presume he also tried to appease possible critics and avoid usual denunciations by striking a "positive" note, be it even by resorting to empty and misleading rhetoric, and finding it necessary to publicly subscribe to the totally trivial thesis of "modifiability in principle."

In fact, it is not the only occasion where Plomin has tried to enhance the public standing of behavior genetics by making dubious but possibly expedient statements, against his better judgment. In an attempt to convince psychiatrists about the usefulness of behavior genetics, Rutter and Plomin (1997: 210) severely criticize Jensen (1969b) for making two elementary mistakes about heritability: (a) the identification of heritability with ineffectiveness of environmental interventions and (b) the fallacious jump from within-group heritability to between-group heritability. It is undeniable that in approaching a wider public it may be helpful, strategically, to dissociate oneself from Jensen's notorious paper – which, as Plomin says elsewhere, made behavior genetics "suffer by association" (Plomin

1988: 111), and "almost brought the field to a halt" (Plomin 2004: 113). But the attribution of these two errors to Jensen has no basis in Jensen's texts. And, of course, Plomin himself knows this very well. For, in other articles, written mainly for the insiders to the field of behavior genetics, Plomin states that Jensen's section on heritability from the very paper in which he allegedly committed the two blunders "is still the best intro- duction to the genetics of intelligence" (Plomin 1987: 41; cf. Plomin 2003, 107). Oyama is right: it is hard to have it both ways.[3]

5.6 LIMITS TO EGALITARIANISM

If there is neither statistical interaction nor statistical correlation between genes and environments, the environmental contribution to phenotypic variance (sometimes called environmentality) is obtained by subtracting heritability from total variation,[4] i.e., $1 - h^2$. Can we then say that, of the two components of variance, the heritable part of variation cannot be decreased by eliminating the existing environmental differences, while the environmentally caused inequalities are a ready target for elimination? The first claim is true,[5] but the second claim requires more clarification.

Behavior geneticists usually distinguish two kinds of environmen- tal influences: *shared* environments, which tend to make siblings liv- ing together more alike (typically things like SES, school quality,

---

[3] By the way, speaking about the Plomin–Oyama exchange, it occasionally appears that Oyama willfully misreads her opponent. For example, she objects that Plomin "even has people responding to children's 'genetic propensities' and 'genetic differences' rather than to the children themselves, as though components of population variance were visible as components of individual phenotypes" (Oyama 1988b). It is hard to believe that she takes this criticism seriously herself. For, it seems rather obvious that what Plomin meant by using that shorthand expression is that people respond to children's behavior, which in turn is a manifestation of a genetic propensity.

[4] Strictly speaking, we should take into account measurement error as well, otherwise it would be included in environmental variance, and as a result heritability would be *under*estimated. As Eysenck remarked (Eysenck 1985: 89), many published heritability figures are actually too low because they are not corrected for attenuation. It is interest- ing that critics of heritability insist so much on the possible presence of factors that, if ignored, tend to overestimate heritability (like G–E interaction and positive G–E corre- lation), while they almost never consider factors that, if ignored, tend to underestimate heritability (like error variance, assortative mating, and negative G–E correlation).

[5] Christopher Jencks claimed that removing the existing environmental differences can actually decrease h^2, but his claim was based on his idiosyncratic definition of "heritable," according to which even the results of racial discrimination would count as "heritable" (see chapter 3).

neighborhood, parental style, etc.), and *non-shared* ones (like birth order, peer group, prenatal and perinatal influences, chance events, etc.). So the environmental variance is partitioned into c^2 (shared differences) and e^2 (non-shared differences). Now if the question about malleability is asked in the sense "To what extent can the existing phenotypic differences be reduced by equalizing environmental influences through *social reforms*?" then it is clear that our focus must be on c^2 (which is a possible target of planned social measures) rather than e^2 (which is connected with environmental influences that are usually much more difficult to manipulate). Therefore, even for the non-heritable differences, it should not be automatically assumed that they are readily removable by environmental interventions. In other words, h^2 puts constraints on modifiability, but this does not mean that malleability reigns free in the remainder part $(1 - h^2)$.

On the contrary, empirical evidence suggests that for many important psychological traits (particularly IQ), the environmental influences that account for phenotypic variation among adults largely belong to the non-shared variety. In particular, adoption studies of genetically unrelated children raised in the same family show that for many traits the adult phenotypic correlation among these children is very close to zero (Plomin et al. 2001: 299–300). This very surprising but consistent result points to the conclusion that we may have greatly overestimated the impact of variation in shared environmental influences.[6] The fact that variation within a normal range does not have much effect was dramatized in the following way by neuroscientist Steve Petersen:

> At a minimum, development really wants to happen. It takes very impov-
> erished environments to interfere with development because the biological
> system has evolved so that the environment alone stimulates development.
> What does this mean? Don't raise your children in a closet, starve them, or
> hit them in the head with a frying pan. (Quoted in Bruer 1999: 188)

But if social reforms are mainly directed at eliminating precisely these between-family inequalities (economic, social, and educational), and if these differences are not so consequential as we thought, then egalitarianism will find a point of resistance not just in genes but also in the

---

[6] There is an important qualification to the conclusion. Adoption studies do not include the worst environments because children are usually not given for adoption to families below the poverty line or with histories of child abuse, drug addiction, etc. There is actually a good reason to believe that drastic environmental deprivation does have a harmful effect. Yet it remains true that, for a number of traits, between-family environmental differences in the normal range (characteristic of approximately 75 percent of the population in the United States) have an almost insignificant impact on phenotypic variation in adulthood.

non-heritable domain, i.e., in those uncontrollable and chaotically emerging environmental differences that by their very nature cannot be an easy object for social manipulation.

All this shows that it is irresponsible to disregard constraints on malleability and fan false hopes about what social or educational reforms can do. As David Rowe said:

> As social scientists, we should be wary of promising more than we are likely to deliver. Physicists do not greet every new perpetual motion machine, created by a basement inventor, with shouts of joy and claims of an endless source of electrical or mechanical power; no, they know the laws of physics would prevent it. (Rowe 1997: 154)

I will end this chapter with another qualification. Although heritability puts constraints on malleability it is, strictly speaking, incorrect to say that the heritable part of phenotypic variance cannot be decreased by environmental manipulation. It is true that if heritability is, say, 80 percent then at most 20 percent of the variation can be eliminated by equalizing environments. But if we consider *redistributing* environments, without necessarily *equalizing* them, a larger portion of variance than 20 percent can be removed.

Table 5.5 gives an illustration how this might work.

In this example with just two genotypes and two environments (equally distributed in the population), the main effect of the genotype on the variation in the trait (say, IQ) is obviously stronger than the environmental effect. Going from G_2 to G_1 increases IQ 20 points, while going from the less favorable environment (E_2) to the more favorable one (E_1) leads to an increase of only 10 points. Heritability is 80 percent, the genetic variance being 100 and the environmental variance being 25. Now if we expose everyone to the more favorable environment (E_1) we will completely remove the environmental variance (25), and the variance in the new population will be 100. The genetic variance survives environmental manipulation unscathed.

But there *is* a way to make an incursion into the "genetic territory." Suppose we expose all those endowed with G_1 to the less favorable

Table 5.5

	E_1	E_2
G_1	115	105
G_2	95	85

environment (E_2) and those with G_2 to the more favorable environment (E_1). In this way we would get rid of the highest and lowest score, and we would be left only with scores of 95 and 105. In terms of variance, we would have succeeded in eliminating 80 percent of variance by manipulating environment, despite heritability being 80 percent.

How is this possible? The answer is in the formula for calculating variance in chapter 1 (see p. 21). One component of variance is genotype–environment correlation, which can have a negative numerical value. This is what has happened in our example. The phenotype-increasing genotype was paired with the phenotype-decreasing environment, and the phenotype-decreasing genotype was paired with the phenotype-increasing environment. This move introduced the *negative* G–E correlation and neutralized the main effects, bringing about a drastic drop in variation.

The strategy calls to mind the famous Kurt Vonnegut story "Harrison Bergeron," where the society intervenes very early and suppresses the mere expression of superior innate abilities by imposing artificial obstacles on gifted individuals. Here is just one short passage from Vonnegut:

> And George, while his intelligence was way above normal, had a little mental-handicap radio in his ear – he was required by law to wear it at all times. It was tuned to a government transmitter and, every twenty seconds or so, the transmitter would send out some sharp noise to keep people like George from taking unfair advantage of their brains. (Vonnegut 1970: 7)

We all get a chill from the nightmare world of "Harrison Bergeron." But in its milder forms the idea that if the less talented cannot be brought up to the level of those better endowed, the latter should then be held back in their development for the sake of equality, is not entirely without adherents. In one of the most carefully argued sociological studies on inequality there is an interesting proposal in that direction, about how to reduce differences in cognitive abilities that are caused by genetic differences:

> A society committed to achieving full cognitive equality would, for example, probably have to exclude genetically advantaged children from school. It might also have to impose other handicaps on them, like denying them access to books and television. Virtually no one thinks cognitive equality worth such a price. Certainly we do not. *But if our goal were simply to reduce cognitive inequality to, say, half its present level, instead of eliminating it entirely, the price might be much lower.* (Jencks et al. 1972: 75–76 – emphasis added)

So although Jencks and his associates concede that excluding genetically advantaged children from school and denying them access to books
may be too drastic, they appear to think that the price of equality could
become acceptable if the goal was lowered and measures made more moderate. Are they suggesting that George keeps the little mental-handicap
radio in his ear but that the noise volume should be set only at half
volume?

6

Science and sensitivity

I ask myself whether the untruth is not better for American
society than the truth.

Nathan Glazer

Men have ratiocination, whereby to multiply one untruth by
another.

Thomas Hobbes

A book on heritability without a chapter on political issues would be
a bit like *Hamlet* without the ghost of Hamlet's father. Even when the
ghost of politics is not addressed at all, it always lurks in the background,
haunting the protagonists and influencing both the tone and dynamics of
the heritability controversy.

But why not exorcise the ghost from the debate once and for all? For, if
we want to understand heritability *as a scientific concept*, is it not advisable
to isolate it from vagaries of political storms that only create confusion,
distrust, and anger? There are two problems with this idea. First, since
in the discussions about heritability, politics has occupied center stage so
forcefully and for so long, we can hardly make sense of what went on if we
neglect such an important element in the story, however irrelevant it may
"objectively" be for the issue at hand. And second, before looking into
these matters more carefully, we cannot actually be sure that heritability
research is indeed devoid of political implications, as many people keep
telling us.

A good way to start this chapter is to consider two frequently used
arguments to cross the barrier between science and politics: first, the claim
that a scientific belief is mistaken because it is politically motivated, and
second, the claim that a scientific belief is politically motivated because it
is mistaken.

6.1 MISTAKEN BECAUSE POLITICALLY MOTIVATED

If somebody accepts a scientific hypothesis merely because it appears to conform to his political opinion, can this fact be used to support the claim that the hypothesis is actually *false*? Such a move looks like a genetic fallacy, an error in reasoning that elementary logic textbooks warn us against: the way a belief is arrived at cannot tell us anything about whether that belief is true or not. But on the other hand, this kind of argument is used all the time. It is a standard practice, for example, to dismiss someone's views about biological differences between human groups once it is shown that these views spring from, say, his right-wing political preconceptions. Is this reasonable?

Writing about the genetic fallacy, Elliott Sober (Sober 1993: 206–207; Sober 1994: 106) argues that the genesis of a belief *can* sometimes point to the falsity of that belief. He gives an example of a teacher who forms a belief about the precise number of students in a class by randomly drawing a slip of paper from an urn. The urn contains a hundred such slips with different numbers written on them. On the basis of drawing the number 78, the teacher starts to believe that there are 78 students in the class. Sober says that in this situation, we are entitled to conclude that his belief is false, i.e., that very probably it is *not* true that there are exactly 78 students in the class. He is right about this, of course, but his justification for the conclusion seems wrong:

> The genetic argument [from the genesis of the belief to its falsity] is convincing. Why? *Because what caused [the teacher] to reach the belief had nothing to do with whether the belief is true*. When this *independence relation* obtains, the genetic argument shows that the belief is implausible. (Sober 1993: 207; cf. Sober 1994: 106)

This sounds odd. For, if the information about the way the belief was caused had *nothing to do with whether the belief is true*, how can it then point to the belief probably being *not true* (i.e., being implausible)? It cannot. If something is not truth-relevant, then by this very fact it is not falsity-relevant either.

So how do we then know, in that situation, that the belief is probably false? Actually, it has nothing to do with the genesis of that belief. The proposition that there are exactly 78 students in the room is most probably false simply because this number is one of many different and equally likely possibilities. Therefore the chance that 78 is the correct number is very small, even *before* anything is known about the genesis of the corresponding belief. When we later learn how the belief was formed (by the number 78 being randomly drawn from an urn) this cannot change our attitude toward that belief at all. It is just that the information about the genesis of the belief gives us no reason to revise our *earlier* probability estimate, which was very low before, and remains very low afterwards.

To show that an epistemically arbitrary way a belief is caused cannot in itself be a good reason to think that the belief is false, consider a slightly changed situation. A teacher knows that the number of students in his class must be either even or odd. Imagine that he forms the belief that the number of students in the class is even by randomly drawing a slip of paper from an urn containing *two* slips of paper. He draws a slip with the word "even," while the other slip is "odd." Sober's condition is satisfied here. The "independence relation" obtains: *what caused the teacher to reach the belief had nothing to do with whether the belief is true.* Nevertheless, it is quite clear that now we would *not* be justified in concluding that the belief "The number of students is even" is false. Since the antecedent probability of that belief is 0.5, the fact that it was caused in a way that had nothing to do with whether the belief is true *cannot* show that it is probably false. To sum up, it *is* a genetic fallacy to infer the falsity of a belief merely from this belief having been arrived at in a way that is independent from its truth. "Independent of truth" means "independent of falsity."

The argument of the type "Mistaken because politically motivated" involves at least three steps: (1) that person X has a particular political attitude A; (2) that A was the main cause of X's adopting a scientific belief B; and (3) on the basis of the two previous steps it is concluded that B is false. It is by no means easy to justify the first two steps, but as our discussion demonstrates, even if (1) and (2) are established, (3) does not follow. Proving that someone's belief about scientific matters has been caused by truth-irrelevant considerations is in itself irrelevant for the truth of that belief.

Apply this to the question of the origins of black–white difference in IQ. There are two possibilities there. Either (a) the difference is entirely caused by environmental differences between the two groups, or (b) at least a part of the difference is explainable by genetic differences. If someone comes to accept (a) in a totally irrational way (say, by rejecting the hereditarian view out of hand and without looking at the evidence at all), this would not give us a good reason to think that what he believes is false. Likewise, if someone advocates (b) only because it agrees with his irrational hatred of other races, we cannot take this as a piece of evidence against (b). We cannot use irrational beliefs to discover the truth by just turning them 180 degrees.

An attempt to do just that comes from Michael Dummett, Oxford professor of logic. He says that "it should seem ... *obvious* that contemporary psychologists in the United States and Britain, advancing the thesis of the hereditary inferiority of Negro intelligence, are . . . reflecting prejudices still widespread in these countries," and that hereditarianism about racial differences and IQ, "so obviously conforming to a palpably powerful prejudice, *can be set aside by any rational judge without further examination*" (Dummett 1981: 296, 298 – italics added). In other words, he suggests that the views in question can (and should) be rejected without looking into the relevant empirical evidence at all!

Although one cannot help feeling a peculiar sort of admiration for Dummett's candor here, the fact that he is ready to defend publicly such an idea is yet another sign that the situation in philosophy has deteriorated to the point that it is necessary to ring the alarm bell. For, it is not only that philosophers of science tend to make judgments about heritability without seriously studying the literature, or without even properly understanding the theories they attack. Now, in addition, a thinker of Dummett's stature openly defends an epistemic norm that legitimizes this awkward way of forming beliefs.

6.2 POLITICALLY MOTIVATED BECAUSE MISTAKEN

A much more common polemical move is to argue that someone's scientific opinion must be politically motivated because it is mistaken. For the argument to go through, the mistake involved has to be a gross, inexcusable error that cries out for an explanation, and then a political prejudice is offered as the best explanation of such a colossal cognitive failure.

Anyone who resorts to this argument has to establish the following four things: (1) that scientist X who is accused of a mistake did actually commit the mistake, (2) that the mistake is a serious blunder, rather than one of those bona fide errors that are expected to happen sporadically in the course of normal scientific work, (3) that X also had a particular political attitude, and (4) that the mistake was really due to the influence of that political attitude.

With all the necessary steps of the argument laid bare, the immensity of the task becomes clear. Let us take a quick look into the more fine-grained structure of that argument by examining, in a very general way, some of the problems that might emerge at its successive inferential stages.

Step (1) looks rather straightforward. It seems that the question whether a mistake has been made or not should be easily settled and agreed upon by everyone. But just recall the cases of Morton's craniological measurements (chapter 1) and Jensen's inference from within-group heritability to between-group heritability (chapter 4), where serious political accusations were launched without any cognitive error actually having been committed in the first place. These are not isolated examples. Evidently, extra caution is needed with the argument "mistaken, therefore politically motivated" because it can break down even in its first step.

In step (2), it is occasionally also quite difficult to decide whether we are dealing with an unexceptional, benign scientific mistake due to human fallibility or a blunder of such magnitude that could justify suspicions about irrational influences from the political domain. Therefore, if there is a lot of disagreement about which of these two judgments is right about a particular error, even those who opt for the blunder diagnosis should concede that, since their own negative verdict is massively disputed by others, it may be premature on that basis to mount grievous accusations about political motives.

Steps (3) and (4) are crucial, but they are particularly difficult to defend because they involve a speculation about the psychology of individual scientists. It is sometimes hard in the heat of the battle to resist a temptation to "unmask" the adversary's political agenda, but typically one knows too little about the opponent to make the whole thing anything more than a crude insinuation. It is not surprising, therefore, that the attribution of a political attitude is rarely based on *independent* and *direct* evidence. Rather, it is usually inferred in an indirect way, by insisting that an error is so egregious that it *must have been* the result of a political bias: ideological prejudice is allegedly the best possible explanation of such an outrageous cognitive gaffe. The assumption here is that the (mistaken) scientific belief B is thought, rightly or wrongly, to advance the (imputed) political attitude A, and it is then suggested that X made the mistake *because* he wanted to advance A.[1]

---

[1] Human tendency to impute political motives *only* on the basis of someone's beliefs seems amazingly strong. In one psychological study (reported in Halpern et al. 1996: 262–263), students were asked to judge motives of a professor who in a lecture about intelligence suggested that genetic factors can explain *some* portion of the racial difference in IQ test scores. Knowing nothing else about that person, nearly one in four students considered this information sufficient to conclude that the professor was guided by racist motivation. In a sense, the result should not actually be too surprising because we encounter this type of "attribution error" quite often in our daily lives.

There are four things that can go wrong with this move. First, it may be that very few people actually think that B advances A. Second, even if many people believe that B advances A, X himself may be unaware of this or disagree about it. Third, even if X believes that B advances A, it may be that as a matter of fact he simply does not support A. And fourth, even if X supports A and also believes that B advances A, it still remains an open question whether he defended B *because of* A. On the one hand, his endorsement of both B and A may not be connected at all. The two commitments may require completely different and independent explanations. On the other hand, it may be that the causal connection is actually running in the opposite direction, i.e., that X defended A because of B, not the other way around. That is, rather than being blinded by a political prejudice into making a scientific mistake, it can turn out that X adopted a political attitude because his (mistaken) scientific belief led him to it. (It is irrelevant to complain here that a move from scientific belief to a political attitude is fallacious. For even if a transition from "is" to "ought" is always illegitimate, some people may engage in this kind of inference simply because they are unaware that it is logically impermissible. Furthermore, as will be explained in section 6.5, adopting a political attitude on the basis of a scientific belief may not be an error after all, despite all the outcry against the "is-to-ought" fallacy.)

Another thing bears emphasis here. Even leading scientists occasionally make blatant and elementary mistakes without any sinister motive being at work. Out of many possible illustrations, here is one example that involves a very fundamental misunderstanding of heritability by a prominent psychologist:

> heritability is higher when there is a great deal of variability in the characteristic of interest. If there is minimal variability – for example, being born with five fingers – heritability will necessarily be low. Thus the heritability of IQ is high when the sample studied has some individuals with very low scores and some with very high scores. Heritability will be low if everyone in the sample has an IQ between 95 and 105. (Kagan 1998: 55)

In contrast to what Kagan says, it is an elementary fact about heritability that it *in no way* depends on the amount of total variability. For instance, even "if everyone in the sample has an IQ between 95 and 105" it does not follow that heritability will necessarily be low. It can still be 100 percent, when all of these (small) IQ differences are caused by genetic differences. Kagan is wrong simply because heritability measures the

188

proportion of variation that is due to genetic causes, and is not affected at all by the "absolute" range of phenotypic variation.

In this case there is no temptation to invoke politics in explaining the error, probably because it does not point in any specific political direction. There is no visible political interest that could be served if the erroneous belief were accepted. But then we should keep an open mind for a possibility of this type of benign mistake occurring even in the cases where the mistake could be exploited for political purposes.

After all these admonitions against overhasty "mistaken-therefore-politically-motivated" inferences, let me stress that I am not arguing that scientific mistakes are never caused by political prejudice. My point is just that in any particular case there is a big gap between the mere existence of a cognitive error (however terrible) and the hypothesis that the error was politically motivated. To cross the gap one needs a lot more information than is usually available, certainly more than just the knowledge that the error occurred and that it is thought to advance a given political standpoint. The best evidence, surely, would be to have the author himself say that his expressed scientific opinion was driven by a political agenda. Although this does not happen often, it is not unheard of. One example:

> We share a commitment to the prospect of the creation of a more socially just – socialist – society. And we recognize that a critical science is an integral part of the struggle to create that society. (Rose et al. 1984: ix)

Well, now the situation does change. If Rose, Lewontin, and Kamin say themselves that they regard their own scientific activity as an integral part of the struggle to create a socialist society, and if in their attempts to understand human behavior they draw inspiration from "revolutionary practitioners" like Chairman Mao (Rose et al. 1984: 76), why shouldn't we take their word for it? So, if we see that they defend a scientific belief that is opposed by most experts in the field but which at the same time coheres with some of their declared ideological preconceptions, we can hardly be blamed for taking seriously the possibility that their minority view is less driven by cognitively relevant reasons and more by their political commitments. For instance, their insistence that the heritability of IQ may be zero "for all we know" (Rose et al. 1984: 116) goes squarely against the consensus of behavior geneticists, yet this 100 percent environmentalism can clearly be appealing to radical egalitarians.

There can be nothing wrong with suspecting a particular influence of politics on science in cases where the general influence of this kind is actually self-advertised. The point, however, is that these cases are quite exceptional. Great caution and restraint are needed in the usual situations where political imputation depends on a number of assumptions that are very difficult to prove.

But the temptation of delivering a political coup de grace proves irresistible even to methodologically sophisticated thinkers. For example, the philosopher Clark Glymour, who is well known for imposing super-rigorous standards for causal inference, is nevertheless quite happy, when criticizing hereditarianism, to derive a causal conclusion on the basis of virtually zero evidence. He states with certainty that parts of *The Bell Curve* are "*deliberately* pseudo-scientific" (Glymour 1997: 261 – italics added), e.g., that the authors' acquisition and analysis of data were merely a rhetorical device intended to make their antecedently formed (political) opinion look more acceptable to others. In other words, the claim is that the evidence Herrnstein and Murray gave did not actually determine their views at all but that it was put forward merely as window-dressing for their real, hidden agenda. How on earth could Glymour substantiate such a grave accusation? In fact he does not even try. No argument whatsoever is offered in support of this speculation about intellectually disreputable causes of Herrnstein and Murray's published views. Such is the unbearable lightness of ideological denunciation.[2]

The most vocal anti-hereditarians are especially "trigger happy" with political imputations. Their derivation of political attitudes from scientific opinions occasionally becomes automatic and mechanical to the point of absurdity, as when Leon Kamin is reported to have said that "the simplest way to discover someone's political leanings is to ask his or her view on genetics" (in Herbert 1997). Even in the opening pages of his main contribution to the IQ debate, Kamin suggests quite directly that an interest in heritability may be a sign of some sinister political intentions: "Patriotism, we have been told, is the last refuge of scoundrels. Psychologists and biologists might consider the possibility that heritability is the first" (Kamin 1974: 3).

---

[2] Christopher Hitchcock also seems to think that in some cases no argument is necessary when making accusations of bad faith. Echoing Glymour, he merely asserts that Herrnstein and Murray "abus[ed] statistical tools to defend prejudged conclusions" (Hitchcock 2003: 342). Besides, Hitchcock described *The Bell Curve* as "infamous," hardly the most appropriate way for a philosopher of science to refer to a book that many leading psychologists actually regard as representing the mainstream view in their field (cf. Gottfredson et al. 1994).

The irony is that those people who see political deformations of science everywhere[3] are sometimes unable to notice the most obvious and blatant cases of political intrusion in science. For example, Gould describes with glee how he and Niles Eldredge found like-minded souls among paleontologists in the Soviet Union, who believed in the "laws of dialectics," spoke of the "transformation of quantity into quality," and basically "support[ed] a model similar to our punctuated equilibria" (Gould 1983: 153). It is astonishing that, after the notorious destruction of Soviet genetics by the Lysenkoist imposition of Marxist dogma, Gould could be so naive as to take seriously the Russian paleontologists' public recitation of ludicrous dialectical formulas, which they themselves most probably detested and ridiculed in private. Contrast Gould's admiration for dialectical materialism with Alexander Solzhenitsyn's first-hand report about the status of that "philosophy" in his famous talk "Words of Warning to America":[4] "[I]n the Soviet Union today, Marxism has fallen to such a low point that it has become a joke, an object of contempt. No serious person in our country today, not even university and high-school students, can talk about Marxism without a smile or a sneer" (Solzhenitsyn 1976: 47).

Interestingly, the eagerness to unearth political motivation goes so far that it is sometimes inferred at second remove, not from what the author says, but from the sources he cites. For instance, in his widely read and praised review of *The Bell Curve*, the Oxford philosopher Alan Ryan writes about Herrnstein and Murray: "It is already becoming clear that the air of dispassionate scientific curiosity that they are at such pains to maintain is at odds with the eccentricity of some of their sources" (Ryan 1995: 20). What makes these sources eccentric?

First, Ryan objects to Herrnstein and Murray for treating J. Philippe Rushton's book *Race, Evolution and Behavior* (Rushton 1995) as the work of a serious scholar. Ryan finds the book "bizarre," but his way of dismissing it is itself bizarre. His contempt for the book is based on summarizing its content with a one-sentence quotation from *Rolling Stone* magazine. Oddly enough, Ryan has no misgivings about rejecting so summarily and highhandedly a work that falls completely outside of his area of academic competence. Oxford professors do not usually reject

---

[3] Even the idea in molecular biology that information always flows in the direction DNA → RNA → protein was "unmasked" as being driven by "ideological concern" (Rose et al. 1984: 60).

[4] The talk was delivered in 1975, a year after Solzhenitsyn was expelled from the Soviet Union, and three years *before* Gould's text was first published in *Natural History*.

theories that do not belong to their field of expertise with a sneer and without argument, especially not if these theories are taken seriously by scholars in the relevant research area. In contrast to Ryan's contemptuous attitude toward Rushton, one of the leading anthropologists thinks that Rushton's work deserves not just "serious attention and respect," but even "congratulations" (Harpending 1995). Of course, many would disagree. The opinions of experts are divided here. My point is that if philosophers decide to enter the fray, they are surely expected to be more than mindless cheerleaders against one side in a very complicated scientific debate.[5]

Ryan's second complaint is that in *The Bell Curve*, Richard Lynn is described as a leading scholar of racial and ethnic differences, and no mention is made of "his fondness for the theories of Nordic superiority." Ryan overlooks one crucially important thing about Rushton and Lynn. Much of their research output, however controversial or "bizarre" it may appear to him, has successfully passed the peer review process and has been published in respectable social science journals, on the recommendation of competent experts and after scrupulous evaluation. Of course, this in itself does not mean that their claims are to be accepted, but it does mean that after such validation it would be irresponsible of any scholar who later wrote about these issues to ignore their views or pretend that they are not part of the relevant literature. Also, as long as they can rely on the internal quality control based on the evaluation of journal editors and referees in pertinent fields, scientists are under no obligation to gather information about political opinions of the authors they cite, nor to report about such matters. It is a chilling idea, indeed, that a scientist X could be accused of sharing Y's political views just because X discussed Y's peer-reviewed article but failed to explicitly distance himself from a political view that some people attribute to Y.

There is some inconsistency here as well. The demand that one should warn the reader about a possible political "background" of one's sources is not applied even-handedly across the whole political spectrum, much as we would expect it from a principled approach. To ask a rhetorical question, would Ryan advocate similar vigilance toward the intrusion of far-left politics and demand that, say, scientists who quote Leon J. Kamin's views on IQ and heredity should always announce to the world that he is a Marxist radical and a former member of the Communist Party?

---

[5] Let me also mention that in the article on race in the Oxford *Encyclopedia of Evolution*, Rushton's book is actually one of the only five references to the relevant literature in the last thirty years (Harpending 2002: 981).

Ian Hacking, in his own review of *The Bell Curve*, "especially recommends Alan Ryan's analysis," and he claims as well that it is "useful" to examine Herrnstein and Murray's sources.[6] Hacking also focused on Richard Lynn, "whose research will strike many readers as questionable" (Hacking 1995). But again, if Hacking is right that "many" readers can so easily spot that there is something wrong with Lynn's research on race and IQ, how come that, for example, the editors of *Nature* overlooked these obvious flaws and published one of his articles (Lynn 1982) on that very topic?

Hacking and Ryan can call some of *The Bell Curve*'s sources "eccentric," "bizarre," or "questionable," but their attempt to establish "guilt by citation" suffers from a fundamental problem: whatever *they* may think about Lynn's and Rushton's views, it is simply undeniable that these views are recognized and taken seriously by experts in the relevant fields. No scholar can afford to ignore these contributions to the literature when discussing human intelligence. Therefore, contrary to what the two philosophers say, it is *failing* to cite Lynn and Rushton that would be objectionable in this context, rather than citing them. For instance, in N. J. Mackintosh's book (1998), which is widely praised as a scrupulous and reliable overview of research on human cognitive abilities, Lynn is one of the most frequently mentioned authors.[7] Should we conclude from this fact (by using Ryan's argument) that "the air of dispassionate scientific curiosity" that Mackintosh is "at such pains to maintain is at odds with the eccentricity of some of his sources"? Let us hope not.

6.3 CONSEQUENTIAL FALLACY

Politics can enter discussions about science (behavior genetics, in our case) through another avenue: not by focusing on causes (motivation) but on *consequences*. Here the idea is that envisaged consequences of the possible acceptance (or rejection) of a given scientific theory should be an input in deciding whether that theory should be accepted (or rejected). This reasoning is sometimes called the "consequential fallacy."[8]

---

[6] Similarly, the only publication to which Peter Singer refers the reader for a critique of *The Bell Curve* is a paper about its "tainted sources" (Singer 1996: 229).

[7] Actually no author has more references in Mackintosh's bibliography than Lynn. (And yes, Rushton is quoted there too.)

[8] "Closely related to accepting or rejecting a claim because of its source (rather than because of evidence) is accepting or rejecting a claim because of the harm or good that might be

Philip Kitcher defends it with the following argument in *Vaulting Ambition*:[9]

> Everybody ought to agree that, *given sufficient evidence* for some hypothesis about humans, we should accept that hypothesis whatever its political implications. But the question of what counts as sufficient evidence is not independent of the political consequences. If the costs of being wrong are sufficiently high, then it is reasonable and responsible to ask for more evidence than is demanded in situations where mistakes are relatively innocuous. (Kitcher 1985: 9)

Kitcher also defended the same position more recently:

> where the stakes are high we must *demand more* of those who claim to resolve the issue. So, it was relevant to point to the *political* consequences of *accepting* some of Wilson's claims about human nature because doing so makes us aware of the need for *more rigorous arguments* and *greater certainty* in this area. (Kitcher 1997: 280 – italics supplied)

Note that Kitcher talks about *acceptance* of theories, not just about using or applying theories, or about acting on the assumption that a given theory is true (or false), where the expected utility approach would indeed be perfectly appropriate. His concern is not about application but about acceptance.[10]

What does Kitcher mean by "acceptance"? As Arthur Fine said, "*acceptance* is a specification-hungry concept" (Fine 1990: 109). It is "nicely ambiguous, allowing for various specifications (accept as true, as useful, as expedient, for the nonce, for a *reductio*, etc.)" (Fine 1991: 87). There are three reasons to think that by "accept" Kitcher actually meant "regard as true." First, since this is a commonsense interpretation of "theory acceptance" (cf. Worrall 2000: 349) and since Kitcher did not explain

---

caused by holding the belief (its consequences). This fallacy has no special name (the *consequential fallacy* has been suggested) but it can occur whenever our desires intrude on our reasons for belief " (Salmon 1995: 187).

[9] Although this book is an attack on sociobiology, it is interesting that the theory that Kitcher picks out in the stage-setting, melodramatic first chapter to illustrate political dangers of hereditarianism is not sociobiology but heritability of IQ (see section 1.5.2). As others have also noted: "Sir Cyril Burt's name is used to give a foul smell to research quite unconnected to his own; for instance, by Kitcher in his assessment of 'pop' and human sociobiology" (Caryl & Deary 1996).

[10] The idea that politics should shape epistemic standards is also defended by Anne Fausto Sterling: "I impose the highest standards of proof, for example, on claims about biological inequality, my high standards stemming directly from my philosophical and political beliefs in equality" (Fausto Sterling 1992: 11–12).

the meaning of that term, probably the commonsense connotation was intended. Second, since his book *Vaulting Ambition* was written as much for philosophers as for scientists and the general reader, it is unlikely that he used the term "acceptance" in some esoteric philosophical sense. Third, and most importantly, Kitcher himself connects acceptance with truth when, in arguing for the need to look at political implications when deliberating about acceptance, he says that a theory choice *cannot be just about truth*, pure and simple: "Lady Bracknell's reminder is apposite – the truth is rarely pure and never simple" (Kitcher 1985: 9).[11]

Kitcher's basic worry is that the fact of scientists *merely accepting* a given hypothesis (regarding it as true) can have bad social and political effects. For, it is on the authority of science that the wider public might also accept that hypothesis, and through a presumably complicated process it might all lead to some very bad outcome. Should we follow Kitcher and decide to raise the hurdle for accepting "dangerous" theories?

No, for two reasons: (1) it would lead to irrational behavior, and (2) it would not achieve its purpose.

6.3.1 Irrationalism with a human face

The target of our cognitive efforts (in science and otherwise) is truth. That is, by using fallible truth indicators we try to judge whether H is true or false, and the way we normally function is that only something that we regard as *relevant evidence* can sway us into accepting or rejecting H. Kitcher's recommendation that political considerations should play a role in determining when evidence is "sufficiently strong" for acceptance is a recipe for epistemic irrationality. He exhorts us to "over-believe" theories with beneficial political consequences, and "under-believe" theories with harmful political consequences. (The terms "over-belief" and "under-belief" come from Haack 1996: 60.) The result is that the fine Humean advice that "a wise man proportions his belief to the evidence" is thereby being replaced with the advice that "a wise man proportions his belief (at least in part) to the envisaged consequences of his belief." It is interesting that the Kitcherian consequence-based approach to the evaluation of theories was very common in Hume's time, but then the main concern was over religious and moral repercussions rather than political consequences. Hume condemned that approach in clear terms: "There is no

---

[11] Not that it matters much, but it was not Lady Bracknell who said that. It was her nephew Algernon.

195

method of reasoning more common, and yet none more blamable, than, in philosophical disputes, to endeavour the refutation of any hypothesis, by a pretence of its dangerous consequences" (Hume 1999: 160).

Of course, Kitcher does not advocate the idea that the acceptance of scientific theories should always or exclusively be shaped by their consequences. Yet to the extent that he does so, he indeed urges scientists to behave irrationally. Since his proposal is inspired by good intentions it can be called "irrationalism with a human face." Take a scientist who accepted H on the basis of what he regarded as adequate reasons. Suppose that he realizes later that his accepting H could have some unwanted social consequences. Is it not glaringly obvious that, as long as he remained epistemically rational, he simply *could not* reject H *just because* this would be more beneficial, politically?

One can imagine extreme situations in which one would look at envisaged political consequences in deliberating whether to *publicly defend* certain scientific views.[12] But Kitcher is not talking about this. He is talking about *accepting* these views. To see how radical his claim is, consider a non-scientific example. I accept that Oswald killed Kennedy. Now suppose that I am told, credibly, that something horrible will happen to me if I continue accepting the Oswald theory and if it turns out that the theory is false. Obviously, under the circumstances it would be in my strong interest to apply higher standards of evidence to my opinion, in the hope that in this way I will be able to get rid of it, and thereby avoid the danger of horrible consequences. But how could I do it? As much as I might wish to become a "skeptic" about the Kennedy assassination, I simply could not transform myself into one, merely on the account of the possible consequences. True, if given enough time to weave a web of self-deception, I could manipulate my own cognitive abilities (by self-indoctrination, hypnosis, brainwashing, etc.) and I could eventually end up being unsure whether it was Oswald or Lyndon Johnson who killed Kennedy, but this just illustrates that the project can only be executed in an irrational way.

Another misconception that clouds the issues is the idea that if I accept H (i.e., regard H as true), then this automatically implies that, in any situation, I will behave as if H is true. This is wrong. Acceptance is not

---

[12] In my opinion, such a decision would be justified only under very exceptional circumstances. For example, a scientist living under a totalitarian regime would have a moral obligation not to defend certain views that could be grist for the mill of the oppressive ideology (mainly because he would not have an opportunity to explain the true meaning of his opinion). I think that this kind of scientific self-censorship in theory choice is almost never called for in liberal-democratic countries.

so rigidly linked to action. For example, I do accept the Oswald theory about Kennedy's assassination, but I would certainly not bet the life of my daughter on it. Similarly, if scientists accept H, this means that they think H is much better empirically supported than its rivals, that the evidence for H is very strong, etc., but it does *not* mean that they are advising people to behave as if H is true on all occasions. After one hears the scientific opinion, even if one trusts it completely, one still has to use one's own judgment in deciding what to do. For example, if geologists say that they believe that no big earthquake will hit Texas in the near future, this makes it reasonable for Texans not to buy earthquake insurance for their homes, and in general to go about their lives as if a big quake will never happen. But it certainly does *not* make it reasonable for people who build nuclear power plants in Texas to behave in the same way. They also probably trust geological science, yet it is their duty to prepare for the eventuality that the *accepted* scientific prediction is wrong. Kitcher's idea that scientists should accept hypotheses depending on the consequences sounds quite odd here. Wouldn't it be strange if geologists informed the general public that Texas is in the seismic zone 0 (i.e., at virtually no risk of damage from earthquakes), but if they then told the nuclear power plant builders that, given the more serious consequences of a mistake in this context, the "zone 0 hypothesis" can no longer be accepted?

What makes Kitcher's advice especially extravagant is his insistence (Kitcher 1985: 9) that the rationality of adopting a scientific hypothesis should depend not only on the political costs of *accepting* it, but also on the possible political costs of *failing to accept it*.[13] This means that in cases where the political costs of *not accepting* a certain empirically dubious hypothesis were considered to be sufficiently high, we would be actually goaded to make a politically motivated effort to *accept* that theory, against our better (epistemic) judgment. So apparently Kitcher would have nothing but words of praise for the behavior of scientists who first rejected hypothesis H because they thought it inadequately supported by evidence but who, later, after learning that the rejection of H (or indecision about H) could have harmful social effects, promptly lowered their critical standards and obligingly accepted H.

---

[13] Stranger still, he argues that acceptance of a theory should depend *also* on the costs and benefits of adopting it, if it is true. This seems to go against his previous claim that any theory should be accepted, given sufficient evidence. For, if the costs of adopting a theory, if it is true, become prohibitively high, why shouldn't this be a reason, on Kitcher's own logic, not to accept that theory even when we have the best possible evidence that it is true?

Here is another odd implication of Kitcher's view. Suppose that scientists had to choose between two rival hypotheses, H_1 (bad consequences if accepted-but-false) and H_2 (no bad consequences if accepted-but-false). Suppose also that the two hypotheses have exactly the same degree of empirical support (before the introduction of political considerations). Now, in this situation the scientists who took to heart the lesson about politically responsible science would gerrymander methodological standards in the Kitcherian spirit (by holding H_1 to "higher standards of evidence") and they would then declare that they are closer to accepting H_2 than H_1 – *although they would not be able to point to any theoretical or empirical advantage of H_2 over H_1*!

An additional problem arises because the degree of confirmation of a hypothesis would become oddly context dependent. Take a "dangerous" hypothesis H_1, which is initially not well supported by evidence. Imagine now that we notice that accepting another highly confirmed hypothesis H_2, which belongs to a politically neutral research area, would make H_1 itself strongly confirmed. It seems that Kitcher's imperative of political responsibility would require that we hold the "neutral" hypothesis H_2 to higher standards of evidence, because under the circumstances a too easy acceptance of H_2 would do social harm in an indirect way, by vindicating H_1. So political considerations would urge us to assign different degrees of confirmation to the hypothesis in different contexts although the hypothesis would *in itself* have no political content whatsoever.

Implementing Kitcher's proposal would inevitably lead to the high politicization of scientific discussions. Scientists are likely to differ widely in their judgments about the degree of political danger of different hypotheses. For example, if the political Pandora's box is opened, socialists would probably want to apply higher standards of acceptance to theories that they see as threatening to *their* ideology, whereas conservatives would disagree, saying that they envisage the possible truth of these theories with equanimity (or with glee!), and that they see no reason for panic or any kind of "protective measures." With other theories, the roles may be reversed. In any case, the evaluation of evidence in theoretical conflicts would no longer be possible strictly on the empirical merits of rival theories. Sectarian politics would become an integral part of science. (Of course, we know that in fact political considerations are already influencing scientific discussions in some contexts, but at present scientists at least have a bad conscience if they realize that such "external" factors determine their views. Kitcher would help them feel pride, instead of guilt.)

Finally, if the politically inspired tightening of standards of criticism is going thus to increase in proportion to the degree of political hazard ascribed to scientific theories, then with respect to some hypotheses containing what some see as "very dangerous knowledge," methodological requirements for acceptance would at some point become so tough that, given the notorious fallibility of human judgment and the essentially conjectural nature of all science, these theories would, as a matter of fact, be put effectively and forever beyond our ken.[14] Is this political sensitivity or dogmatic rejection, or both?

All this shows that Kitcher's politically calibrated methodology of science inevitably leads to the intellectual corruption and degradation of science. Kitcher's proposal to introduce higher standards of evidence seems to be already in place in what Linda Gottfredson calls one-party science: "the disfavored line of work is subjective to intensive scrutiny and nearly impossible standards, while the favored line of work is held to lax standards in which flaws are overlooked" (Gottfredson 1994: 57).

6.3.2 ... and it is self-defeating too

Let's forget about irrationality and ask whether Kitcher's proposal would be effective. To answer this question, first note that if politics-laden methodological standards are to be adopted, they have to be publicly advertised as an addition to the existing "norms of science." So, if the new "politically concerned" science became widespread, everyone would know that scientists tend to downgrade the "real" plausibility of politically sensitive theories. It does not matter whether scientists would actually manage sincerely to believe that these theories are less plausible, or whether they would just lie about it (for the sake of the good effects of that "noble lie"). The important point is that, since the strategy would be a matter of public knowledge, for this very reason it couldn't work.

---

[14] Some scholars think that in his criticism of sociobiology Kitcher already raised the methodological bar so high that not only human sociobiology but much evolutionary biology would fail to meet his standards, and that "the life sciences simply could not proceed at all" if they had to satisfy Kitcher's unrealistic demands (Rosenberg 1987: 80). This is probably one of the main reasons why E. O. Wilson didn't think it worthwhile to respond to Kitcher's book. But Wilson must have also been put off by Kitcher's mixing of science and politics, which he (Wilson) has always regarded as opprobrious: "Opprobrium, in my opinion, is deserved by those who politicize scientific research, who argue the merits of analysis according to its social implications rather than its truth" (Wilson 1977).

For example, if you learn that, according to the new rule of the game, a scientific theory is pronounced less acceptable in proportion to the perceived political danger of its acceptance, you will immediately realize that what is happening here is that scientists just started using a kind of coded language, and that you have to do some translating to get at the true meaning of their statements. Roughly, if scientists declared that the genetic explanation of the racial IQ difference is "extremely implausible," you would have to take into account that they probably "marked down" the confirmation degree of that hypothesis because of its possible political dangers and you would accordingly add a "correction factor" and interpret them as *really* saying that the hypothesis "has some empirical support." In a similar vein, the phrase "has some empirical support" would be understood to mean "plausible," and "plausible" would be translated as "proved."[15]

So, Kitcher's proposal would achieve nothing. Strictly speaking, this is not quite true. It would actually have a perverse effect of making some politically dangerous theories look *more plausible* than they really are. For instance, take one of these theories, H, that is extremely implausible (judged strictly by empirical evidence, i.e., before its political implications are even considered). Now if all social scientists stated (correctly) that H is extremely implausible, in the world regulated by Kitcher's norms they would simply be unable to convey the meaning of that statement to the public. To see why, imagine, for the sake of simplicity, that they use the seven-point scale in Table 6.1 to rank scientific theories according to their plausibility.

Now look what will happen if the scientific consensus puts H (a politically dangerous hypothesis) in category 7 (extremely implausible). Since everyone would know that dangerous theories tend to be shifted

---

[15] Far from being merely a philosopher's suggestion, it sometimes really happens that words in science change their meaning with political context. For example, in the report "Genetics and Human Behavior" issued by the Nuffield Council on Bioethics, we read that "black individuals score on IQ tests *slightly lower* than white individuals" (Nuffield 2002: 69, italics supplied), and then on the same page we read that the rise in average IQ in Great Britain since World War II (due to the Flynn effect) was "*particularly great.*" But what is the difference between "particularly great" and "slight" here? Surprisingly, in terms of the number of IQ points it turns out that there is no difference at all. If we compare the rise in IQ in Great Britain with the measured IQ difference between whites and sub-Saharan blacks we are talking about a difference of essentially the same magnitude (about two standard deviations). So it is one and the *same* quantitative difference that in a neutral context (Flynn effect) comes to be called "particularly great," and in the politically sensitive context (race) becomes de-emphasized and categorized as "slight." Is it likely that this game of semantic hide-and-seek will do any good?

Table 6.1

Rank	Degree of confirmation
1	Proved
2	Probably true
3	Plausible
4	As likely to be true as false
5	Promising
6	Implausible but worth exploring
7	Extremely implausible

downward on the scale (in order to make them harder to accept and thereby avoid possible bad consequences), it would not be clear whether H received the worst evaluation because (a) it is really in such a bad shape (epistemically), or because (b) in reality it belongs to category 6 or 5, but it was downgraded merely for political reasons.

All this shows that Kitcher's proposal is not only irrational and ineffective but self-defeating as well. Many people think that the most effective weapon against social prejudice (for example, racism) is to show that the theories invoked in its support are rejected by science. But as we just saw, mixing politics with methodology would seriously undermine the credibility of this very verdict. Namely, since the "downward shift" reaches the rock bottom with judgments like "unscientific" or "pseudo-scientific," then if politically dangerous theories were known to be judged more harshly than they deserve on purely epistemic grounds, this practice would actually make it quite reasonable, under the circumstances, to take the dismissal of these theories with a large helping of salt. In other words, if scientists from the relevant fields collectively issued a statement that the theories used to support racism are preposterously false, people would be perfectly entitled not to believe them.

6.4 DOUBLE STANDARDS

The truly bad news is that the politically inspired double standard in science is not just a philosopher's idea. It is already endorsed and practiced by many scientists. However, in contrast to Kitcher's radical proposal that aimed at a politically motivated transformation of *cognitive attitudes* (acceptance of theories), scientists take a more pragmatic stance: if they let politics in, they typically argue that potentially dangerous theories should be contained by simply making it more difficult for that kind of research to be funded, published, publicly presented, etc. Here is a typical proclamation in that vein:

> stricter criteria for the performance and evaluation of studies in behavioral genetics should be implemented. This requires a special awareness by the granting agencies, journal editors, and referees of both the technical issues involved and the potential social consequences of misinformation. (Beckwith & Alper 2002: 325)

A good illustration that the advice has already been implemented is the way an editor of the *American Psychologist* (the flagship journal of the American Psychological Association) justified rejecting Jensen's paper on the explanation of the black–white difference in IQ: "My own feeling as Editor is that since the area is so controversial and important to our society, I should not accept any manuscript that is less than absolutely impeccable" (quoted in Gottfredson 2005). Well, this comes perilously close to outright censorship. For, indeed, if the condition of acceptance is that the manuscript be "absolutely impeccable," who can throw the first stone?

Of course, explicit requests for censorship are rare but they are not unheard of. For instance, Hilary Putnam proposed in 1971 that the American Philosophical Association join the American Anthropological Association in *condemning the publication* of theories of Herrnstein, Shockley, and Jensen, "especially in view of the destructive political uses to which such views are put" (*Proceedings and Addresses of the APA*, vol. 45, 1971–1972, pp. 172–173). The proposal was put to mail ballot but I was unable to find out about the voting result. I contacted Putnam himself and some APA officers from that period with a request for more information, but no one seems to remember how the whole episode ended.

To mention another high-profile intervention in that spirit, while Donald Campbell was President of the American Psychological Association in 1975 he urged members at an annual convention to do "plenty of hissing and booing" at Jensen's invited address on test bias (Gottfredson 2005). Naturally, when professors boo their colleagues, this is a signal to students to express dissatisfaction in their own ways. What Jensen and Herrnstein experienced in the early 1970s is described in detail in Jensen 1972a and Herrnstein 1973, but it is interesting that even Sandra Scarr, with her moderate views, was not left alone. After one public discussion about behavior genetics at her university she was told by students "to get out of town or [she] would be killed" (Scarr 1987: 226).

For a recent example of how far even established scholars are still prepared to condone deliberate distortions of "dangerous" theories "for the
public good," here is what Daniel Dennett had to say about misinterpretations of Rose, Lewontin, and Kamin (1984) and their often unfair
criticisms of hereditarians:

> I don't challenge the critics' motives or their tactics; if I encountered people
> conveying a message I thought was so dangerous that I could not risk giving
> it a fair hearing, I would be at least strongly tempted to misrepresent it, to
> caricature it for the public good. I'd want to make up some good epithets,
> such as genetic determinist or reductionist or Darwinian fundamentalist,
> and then flail those straw men as hard as I could. As the saying goes, it's
> a dirty job, but somebody's got to do it. Where I think they go wrong is
> in lumping the responsible cautious naturalists (like Crick and Watson,
> E. O. Wilson, Richard Dawkins, Steven Pinker, and myself) in with the few
> reckless overstaters and foisting views on us that we have been careful to
> disavow and criticize. (Dennett 2003b: 19–20)

So, Dennett would be *at least* strongly tempted to do "the dirty job"
himself! Notice that he actually does *not* say he wouldn't do it. The only
thing he unequivocally objects to is that *he* be on the receiving end of
this kind of vilification.[16] Conspicuously missing on his list of good guys
("responsible cautious naturalists") are the names of people like Jensen,
Herrnstein, and Eysenck – the authors who were notoriously most often
and most vehemently attacked for conveying a "dangerous message."
Could it be that *they* are actually those unnamed "reckless overstaters"
and "evil guys" (Dennett's term, too), who supposedly deserve what is
coming to them? In fact, it is really hard to see whom else Dennett could
have had in mind with these "good epithets," especially among those
contemporary scholars that were attacked in Rose et al. 1984. Therefore,
there is a very good reason to think that what he wanted to say is, simply,
that the hatchet treatment is OK for people like Jensen, Herrnstein, and
Eysenck, but not OK for people like Dennett, Watson, and Crick.

Richard Dawkins uses a similar strategy to defend himself from the
attack of the same "fire brigade,"[17] whose most active members are Rose,

---

[16] Dennett does say that the strategy he describes is "dishonest," but he also explicitly says
that he does *not* challenge it.

[17] "Critics of biological determinism are like members of a fire brigade, constantly being
called in the middle of the night to put out the latest conflagration, always responding to
immediate emergencies, but never with the leisure to draw up plans for a truly foolproof
building. Now it is IQ and race, now criminal genes, now the biological inferiority of
women, now the genetic fixity of human nature. All of these deterministic fires need to

Lewontin, and Kamin. After demolishing some of the firefighters' central arguments and after making fun of their absurd politicization of science, Dawkins concludes in the penultimate paragraph of his review that the book *Not in Our Genes* is "silly, pretentious, obscurantist and mendacious" (Dawkins 1985). But then, surprisingly, in the last paragraph he bends over backwards, like "decent liberal people who will simply bend over backwards to be nice to *anyone* attacking racialism and Cyril Burt," and says that the "chapters, presumably by Kamin, on IQ testing and similar topics, do partially redeem this otherwise fatuous book." Again, the message is clear: when the "dialectical biology" approach is used against Dawkins or Wilson it is silly, pretentious, fatuous, obscurantist, and mendacious, but when it is directed against Jensen or Herrnstein it somehow miraculously becomes plausible and even praiseworthy.[18] Notice also the not too subtle suggestion that, among those who emphasize the importance of genetics for social science, one should distinguish between "decent liberal people" (e.g., Dawkins himself) and racists (e.g., certain not explicitly identified researchers "on IQ testing and similar topics").

David Lykken had a good comment on this tendency of some Darwinians (he had John Tooby and Leda Cosmides in mind) to publicly dissociate themselves from behavior genetics, in the hope that this move would make their own research less vulnerable to political criticisms: "Are these folks just being politic, just claiming only the minimum they need to pursue their own agenda while leaving the behavior geneticists to contend with the main armies of political correctness?" (Lykken 1998b).

There are some obvious, and other less obvious, consequences of politically inspired, vituperative attacks on a given hypothesis H. On the obvious side, many scientists who believe that H is true will be reluctant to say so, many will publicly condemn it in order to eliminate suspicion that they might support it, anonymous polls of scientists' opinions will give a different picture from the most vocal and most frequent public pronouncements (Snyderman & Rothman 1988), it will be difficult to get funding for research on "sensitive" topics,[19] the whole research area will

---

be doused with the cold water of reason before the entire intellectual neighborhood is in flames" (Rose et al. 1984: 265).

[18] Actually, the "redeeming" chapter on IQ contains most of the fallacies and confusions about heritability discussed in this book.

[19] In the preface of their important and well-argued book *Race Differences in Intelligence*, Loehlin, Lindzey, and Spuhler say that their search for project funds "proved to be quite elusive" both with federal agencies and private foundations, and they add that some

be avoided by many because one could not be sure to end up with the "right" conclusion,[20] texts insufficiently critical of "condemned" views will not be accepted for publication,[21] etc.

On the less obvious side, a nasty campaign against H could have the unintended effect of *strengthening* H epistemically, and making the criticism of H look less convincing. Simply, if you happen to believe that H is true and if you also know that opponents of H will be strongly tempted to "play dirty," that they will be eager to seize upon your smallest mistake, blow it out of all proportion, and label you with Dennett's "good epithets," with a number of personal attacks thrown in for good measure, then if you still want to advocate H, you will surely take extreme care to present your argument in the strongest possible form. In the inhospitable environment for your views, you will be aware that any major error is a liability that you can hardly afford, because it will more likely be regarded as a reflection of your sinister political intentions than as a sign of your fallibility. The last thing one wants in this situation is the disastrous combination of being politically denounced (say, as a "racist") *and* being proved to be seriously wrong about science. Therefore, in the attempt to make themselves as little vulnerable as possible to attacks they can expect from their uncharitable and strident critics, those who defend H will tread very cautiously and try to build a very solid case before committing themselves publicly. As a result, the quality of their argument will tend to rise, if the subject matter allows it.[22]

---

foundations "even suggested that they would only be interested if we were willing to specify in advance just what the conclusions of the study will be!" (Loehlin et al. 1975: ix).

[20] A dramatic example is again Sandra Scarr. She said: "As my friends know, I was prepared to emigrate if the blood-grouping study had shown a substantial relationship between African ancestry and low intellectual skills. I had decided that I could not endure what Jensen had experienced at the hands of colleagues" (Scarr 1988: 56–57). Fortunately for Scarr, her research results turned out to be politically favorable. But it would be interesting to know where she actually planned to flee to escape the rage of her colleagues, if worse came to worse. (Not many good choices there.)

[21] In the Jensen uproar in 1969, the editors of the *Harvard Educational Review* commissioned an article about the whole affair from philosopher Michael Scriven, but then in the atmosphere where attacking Jensen was the order of the day they refused to publish it because it was not sufficiently critical. The paper came out in another journal later (Scriven 1970), but although it was one of the most thoughtful contributions to the debate, it was drowned out in a cacophony of much louder noises.

[22] This is not guaranteed, of course. For example, biblical literalists who think that the world was created 6,000 years ago can expect to be ridiculed as irrational, ignorant fanatics. So, if they go public, it is in their strong interest to use arguments that are not silly, but the position they have chosen to advocate leaves them with no good options. (I assume that

It is different with those who attack H. They are regarded as being on the "right" side (in the moral sense), and the arguments they offer will typically get a fair hearing, sometimes probably even a hearing that is "too fair." Many a potential critic will feel that, despite seeing some weaknesses in their arguments, he doesn't really want to point them out publicly or make much of them because this way, he might reason, he would just play into the hands of "racists" and "right-wing ideologues" that he and most of his colleagues abhor.[23] Consequently, someone who opposes H can expect to be rewarded with being patted on the back for a good political attitude, while his possible cognitive errors will go unnoticed or unmentioned or at most mildly criticized.

Now, given that an advocate of H and an opponent of H find themselves in such different positions, who of the two will have more incentive to invest a lot of time and hard work to present the strongest possible defense of his views? The question answers itself. In the academic jungle, as elsewhere, it is the one who anticipates trouble who will spare no effort to be maximally prepared for the confrontation.

If I am right, the pressure of political correctness would thus tend to result, ironically, in politically *in*correct theories becoming better developed, more carefully articulated, and more successful in coping with objections. On the other hand, I would predict that a theory with a lot of political support would typically have a number of scholars flocking to its defense with poorly thought out arguments and with speedily generated but fallacious "refutations" of the opposing view.[24] This would explain why, as Ronald Fisher said, "the best causes tend to attract to their support the worst arguments" (Fisher 1959: 31).

Example? Well, the best example I can think of is the state of the debate about heritability. Obviously, the hypothesis of high heritability of human psychological variation – and especially the between-group heritability of IQ differences – is one of the most politically sensitive topics in contemporary social science. The strong presence of political considerations in this controversy is undeniable, and there is no doubt about which way the political wind is blowing. When we turn to discussions in

---

it is not a good option to suggest, along the lines of Philip Gosse's famous account, that the world was created recently, but with the false traces of its evolutionary history that never happened.)

[23] A pressure in the opposite direction would not have much force. It is notorious that in the humanities and social science departments, conservative and other right-of-center views are seriously under-represented (cf. Ladd & Lipset 1975; Redding 2001).

[24] I am speaking of tendencies here, of course. There would be good and bad arguments on both sides.

this context that are ostensibly about purely scientific issues two things are striking. First, as shown in previous chapters, critics of heritability very often rely on very general, methodological arguments in their attempts to show that heritability values cannot be determined, are intrinsically misleading, are low, are irrelevant, etc. Second, these global methodological arguments – although defended by some leading biologists, psychologists, and philosophers of science – are surprisingly weak and unconvincing. Yet they continue to be massively accepted, hailed as the best approach to the nature–nurture issue, and further transmitted, often with no detailed analysis or serious reflection.

This kind of situation cries out for explanation. The sheer level of ignorance, distortion, and flawed reasoning that characterizes the "anti-heritability" camp is unprecedented in science and philosophy of science. Could it be that, in accordance with the above description, the drastic decline of standards is here due to the dominant intellectual atmosphere, in which those set to undermine heritability can hope to be praised for their political sensitivity and for opposing a dangerous theory, while at the same time they do not have to worry about being severely penalized for possible shortcomings in their logic and methodology? Could this be an explanation?

Notice that, by offering this hypothesis, I am not imputing political bias to anyone in particular. Recall that I myself previously expressed reservations about speculative individual psychology of that kind, which would aim to unearth motives of a given person on the basis of his cognitive lapses (the move "politically motivated because mistaken"). My explanandum is not the behavior of a particular person (or people). Rather, what I am interested in is a curious *pattern* or *trend*, clearly recognizable in the debate about heritability. My idea is simply to try to connect two general facts that represent important aspects of that debate: (1) the anti-hereditarian message is welcome politically, and is generally much preferred over its alternative;[25] (2) there are many who are ready to deliver this wanted message by sidestepping empirical research (with all its unpredictable results), and by attacking hereditarianism with *a priori* arguments that are simple, apparently definitive – and deeply confused. The best way to make sense of (2) is to explain it as being caused by (1). The inference gains in plausibility if we remember that some of the leading

---

[25] For example, the claim that genetically based racial differences in psychology are negligible or non-existent is usually communicated as good news. On the other hand, no prominent academic ever rejoiced in supporting the opposite view.

critics of heritability admit explicitly that they are motivated by political considerations.

6.5 FROM "IS" TO "OUGHT," NON-FALLACIOUSLY

The time is now to address the central issue: the relation between heritability and politics. Is the true relation that there is no relation? The question is intriguing because, in trying to answer it, we are pulled in opposite directions by powerful considerations. On the one hand, Hume's law that "ought" cannot be derived from "is" seems to guarantee in advance that any empirically discovered value of heritability will be unable to push us in any specific political direction. On the other hand, hereditarians about racial differences are often accused of defending that particular view because of its "racist" implications. Which side is right?

First of all, the issue about heritability is obviously a purely empirical and factual one. So there is a strong case for denying that it can affect our normative beliefs. But it is worth noting that the idea that a certain heritability value could have political implications was not only criticized for violating Hume's law, but also for being politically dangerous. Bluntly, if the high heritability of IQ differences between races really has racist implications then it would seem that, after all, science could actually discover that racism is true.

The danger was clearly recognized by David Horowitz in his comments on a statement on race that the Genetics Society of America (GSA) wanted to issue in 1975. A committee preparing the statement took the line that racism is best fought by demonstrating that racists' belief in the heritability of the black–white difference in IQ is disproved by science. Horowitz objected:

> The proposed statement is weak morally, for the following reason: Racists assert that blacks are genetically inferior in I.Q. and therefore need not be treated as equals. The proposed statement disputes the premise of the assertion, but not the logic of the conclusion. It does not perceive that the premise, while it may be mistaken, is not by itself racist: it is the conclusion drawn (wrongly) from it that is racist. Even if the premise were correct, the conclusion would not be justified . . . Yet the proposed statement directs its main fire at the premise, and by so doing seems to accept the racist logic. It places itself in a morally vulnerable position, for if, at some future time, that the premise is correct, then the whole GSA case collapses, together with its case for equal opportunity. (Quoted in Provine 1986: 880)

The same argument was made by others:

> To rest the case for equal treatment of national or racial minorities on the assertion that they do not differ from other men is implicitly to admit that factual inequality would justify unequal treatment. (Hayek 1960: 86)

> But to fear research on genetic racial differences, or the possible existence of a biological basis for differences in abilities, is, in a sense, to grant the racist's assumption: that if it should be established beyond reasonable doubt that there are biological or genetically conditioned differences in mental abilities among individuals or groups, then we are justified in oppressing or exploiting those who are most limited in genetic endowment. This is, of course, a complete *non sequitur*. (Jensen 1972a: 329)

> If someone defends racial discrimination on the grounds of genetic differences between races, it is more prudent to attack the logic of his argument than to accept the argument and deny any differences. The latter stance can leave one in an extremely awkward position if such a difference is subsequently shown to exist. (Loehlin et al. 1975: 240)

> But it is a dangerous mistake to premise the moral equality of human beings on biological similarity because dissimilarity, once revealed, then becomes an argument for moral inequality. (Edwards 2003: 801)

To see how dangerous it can be to leave the acceptability of a political view to the vagaries of science, consider the following "scientific refutation of racism," defended by Dobzhansky in one of his less guarded moments: "Two basic facts refute the racists: the broad overlap of the variation curves for IQs and other human abilities, and the universal educability, and hence capacity for improvement, however that be defined" (1973: 91). About the first "fact," *how broad* should the overlap of the IQ curves actually be for the refutation to work? Unfortunately, in some situations the overlap does not seem to be broad by any measure. For example, according to some studies of the SAT score distributions of American undergraduates, the overlap between the group comprised of blacks and Hispanics and the group consisting of whites and Asians is only 10 percent. This means that 90 percent of those in the first group have a lower SAT score than 90 percent of those in the second group, leading thus virtually to a bimodal statistical distribution of scores (Jensen 1991). If the numbers are correct, then it appears that, on Dobzhansky's logic, racist behavior in that context would become less indefensible.

The second "fact" is even a less felicitous example. If "universal educability" is taken to mean that there are *known* educational measures which, if they could only be applied to everyone, would eliminate, or significantly decrease, the existing differences in IQ and other human abilities, then this claim is simply false. To see how limited the results of best known attempts to raise intelligence are, see Spitz 1986. Later research has also confirmed that the exuberant optimism about omnipotence of education is not justified (Brody 1997; Baumeister & Bacharach 2000).

Another example of an "unpleasant surprise" in using empirical science to fight racism is the fate of Franz Boas's study (Boas 1912) that was widely regarded as showing that even racial differences in cranial morphology were due to environmental causes. The study was a favorite weapon against racism for ninety years, but it was recently challenged by Corey S. Sparks and Richard L. Jantz. On the basis of a careful statistical reassessment of Boas's data they conclude: "Reanalysis of Boas' data not only fails to support his contention that cranial plasticity is a primary source of cranial variation but rather supports what morphologists and morphometricians have known for a long time: most of the variation is genetic variation" (Sparks & Jantz 2002: 14637). This prompted the following reaction of David Thomas, curator of anthropology at the American Museum of Natural History in New York: "I have used Boas's study to fight what I guess could be considered racist approaches to anthropology. I have to say that I am shocked at the findings" (quoted in Wade 2002).

What was shocking is that the "facts" that were used to fight racism in anthropology turned out not to be facts at all. People who cheerfully use science in their political battles are often unaware that they make themselves hostages of empirical fortune. If it is true that "*fortunately* for ordinary moral standards . . . the tenets of racism are not merely unsubstantiated by the facts but in large measure contradicted by the facts" (Cummings 1967: 58 – italics added), then obviously things can change in the future: further research can revise scientific knowledge and deliver the "unfortunate" news.

All this shows that ascribing racist implications to an empirical hypothesis is a double-edged sword because, however unlikely it may be, the probability of that hypothesis is never zero. So if at the end of the day it turns out that the hypothesis is true, what shall we do? All become racists? Almost everyone would say no.[26] All this shows that it is not wise

---

[26] Ullica Segerstrale reports about a surprising exception (Segerstrale 2000: 223).

for anti-hereditarians to push the so-called scientific refutation of racism too far, nor to impute racism to someone merely because of his empirical beliefs. Hereditarians are quick to agree on this, saying that they have always insisted that their research was pure science, uncontaminated by politics. Therefore, it appears that we have something of a "bipartisan" accord, plus Hume's law, confirming that heritability is disconnected from politics. But should we let ourselves be convinced so easily? I think not.

Hume's law is unassailable, but its content should be interpreted carefully. Hume's law says that no normative conclusion can be inferred from factual premises only. It does *not* say that factual premises must always be irrelevant for any normative conclusion. In other words, a normative conclusion can be derived only if there are some normative statements among the premises. But with some normative statements as background assumptions a factual statement *can* lead to a specific normative conclusion. Therefore, there is no *a priori* obstacle for an empirically obtained heritability value having political implications.

Before we turn to the specific issue about heritability, let's try to clarify things further in a very general way. The schematic discussion that follows will be fleshed out with more specific content later. Suppose we have a descriptive belief, D, that together with a normative premise, N, leads to a normative conclusion, C. Imagine also that C cannot be derived from D alone or N alone.[27] Now what if it turns out that D is false? It has been argued that D being false cannot *force* us to abandon C. For although without D we can no longer justify C, we can always replace N with N* (another normative judgment) that, when combined with not-D, will lead to C. This is correct. But the fact that we *can* always refuse to abandon C (by relying on N*, instead on N) does *not* mean that we will always be *ready* to do that, nor that it will be a *reasonable* thing to do. Sometimes we will simply be unable to find a minimally acceptable N* that could be used (together with not-D) to justify C. In that case, the empirical discovery that D is false would, for all that matters, amount to effectively undermining C (a normative conclusion). At other occasions, again, it may happen that some people will cling to C so strongly that, when confronted with the falsity of D, they will decide to accept even an implausible N* if it supports C. Others, however, may not have qualms about rejecting C, and they will then just stay with N and whatever follows from it. Notice

---

[27] For illustration, here is a simple, non-political situation in which both D (descriptive statement) and N (normative statement) are necessary to derive C (another normative statement): D = John is a murderer; N = Murderers should go to jail; C = John should go to jail.

that even in such a situation of divided reaction, the spread of knowledge that D is false would have the net effect of increasing the proportion of those who reject C.

6.6 LOOKING INTO THE ABYSS

Denying the very possibility of any troubling political implications of behavior genetics is a position defended both by many environmentalists (because this way they avoid their political views being threatened by future empirical discoveries) and hereditarians (because this way they depoliticize their research and try to avoid usual ideological accusations). Life and social science would indeed be much simpler if they were right. This explains why, in discussing this issue, some scholars resort to extraordinary logical contortions and obfuscations to erect a wall between genetics and politics. Here is Alan Ryan again:

> Is there an intelligence gap between black and white Americans that no passage of time and no social policy can close? If there were, would anything follow about the social policies a humane society should adopt? The answer seems to be that there is good reason to believe that there is a gap, but no conclusive reason to believe that it is unshrinkable; if there were, it would have a good many implications about the need to balance the search for efficiency against the desire for a more humane social order – but it would not dictate how we struck the balance and it would introduce no moral novelties into the calculation. In particular a belief in the importance of inherited differences in IQ need not encourage apocalyptic conservatism. (Ryan 1995: 25)

According to Ryan, the hereditarian explanation would introduce "no moral novelties into the calculation." The only implication worth mentioning would be "the need to balance the search for efficiency against the desire for a more humane social order." He says that even if hereditarianism is accepted, striving to abolish or decrease social inequalities between the races would remain a legitimate political goal. And he is indeed right that, even under these circumstances, some people could still justify their continuing to pursue the goal of racial equality. What he fails to notice, however, and what is the most important implication in this context, is that other people could actually use the new empirical information to justify their *abandoning* that goal.

To show this, let me start with Anthony Appiah's trenchant observation about this sensitive issue: "the main reason why people currently worry about minorities that fail is that group failure may be evidence of injustice to individuals. That is the respectable reason why there is so much interest in hypotheses, like those of Murray and Herrnstein, that suggest *a different diagnosis*" (Appiah 1996: 130 – italics supplied). Basically, all is said here. What Appiah calls "group failure" *may be evidence* of injustice. But then again it may not. Everything hinges on the explanation of *why* different groups have different positions in society. The fact that some groups are less successful than other groups (e.g., that, proportionally, they have fewer physicians, engineers, lawyers, college graduates, teachers, etc., but more people in jail, on welfare, in menial jobs, etc.) is a datum. From now on I will refer to this datum as "racial inequality."

My basic claim is that racial inequality *in itself* does not constitute a fact that is morally condemnable. Whether it is condemnable depends on the *origin* or *source* of this inequality. Racial inequality springing from source A may be condemnable for one moral reason, while racial inequality springing from source B may be condemnable for a *different* moral reason. Again, racial inequality that arises from source C may not be morally condemnable at all. Needless to say, there can be disagreements about whether racial inequality is morally condemnable or not, under a given explanatory scenario. Nevertheless, it seems that most of us can easily agree about some particular sources of racial inequality that they are condemnable.

The most disturbing source of racial inequality is invidious discrimination. If members of one racial group are singled out for special treatment and systematically denied opportunities to achieve their potential, this is a social injustice that calls for redress. The moral imperative to abolish this kind of unfairness is exceptionally pressing. And obviously, if discrimination is the whole story, its elimination would lead to equality. But if hereditarianism is true, the moral nature of the situation changes dramatically. Contrary to what Ryan says, the truth of hereditarianism would indeed introduce a "moral novelty" of great consequence. Namely, were it proved that racial inequalities are due to biology, i.e., that they are *not* the result of discrimination, the most compelling reason to fight these inequalities would disappear. (Of course, racial inequalities might be partly genetic and partly due to discrimination, in which case eliminating discrimination would decrease these inequalities but could not eliminate them completely.) There might be *other* reasons, of course, that egalitarians could try to use to continue the fight for complete racial equality, but in taking that line they should be aware of three things: (1) that they are thereby switching to a completely new way of defending their

213

political goal, (2) that their momentous argumentative shift is caused by nothing else but the empirical triumph of hereditarianism (which invalidates the discrimination hypothesis),[28] and (3) that they shouldn't be surprised if others find the egalitarianism with this *new* moral justification much less convincing or even unacceptable. Surely, a political movement must expect to lose followers if it replaces the electrifying slogan "Down with racial discrimination!" with a catch-all and hollow phrase like "Fight for a more humane social order!" (the political goal that Ryan mentions).

The whole situation can be described by making use of the above schematic representation of the impact of empirical information on the normative sphere. Again, we have three statements: D = Discrimination ("Social differences between blacks and whites are the result of arbitrary discrimination"); O = Opportunity ("Everyone should have an equal opportunity to realize his or her potential"); E = Egalitarianism ("Social differences between blacks and whites should be eliminated"). D is a descriptive statement (i.e., it is not normative), but O and E are normative statements (they contain the word "should"). Now if you believe D and O, you should support E (because D says that the socially inferior position of blacks is caused by their having been denied equal opportunity, and O says that that everyone should have an equal opportunity). What, however, if you reject D (say, because you think that hereditarianism is true and hence the hypothesis of discrimination is false)? There are two questions here. First, can you, as a reasonable person, still support E? Answer: yes, you can, and in fact you should – provided that you find another moral principle, O*, which you accept and which you can use, in conjunction with not-D (and the background knowledge), to justify E. Second, can you, as a morally good person, abandon E? Answer: yes, you can, and in fact you should – provided that you don't find an acceptable *alternative* normative judgment (O*) that could serve as a justificatory bridge to E.

We learn some important things from exhibiting the structure of the argument in this way. First, if hereditarianism is true and if, say, the whole between-group difference is explained by genetic differences, then the discrimination hypothesis is false. But the discrimination hypothesis is an extremely powerful and almost universally accepted argument for egalitarianism. Ergo, if hereditarianism is true, egalitarianism loses an

---

[28] In the sense that those differences that are heritable are not the result of discrimination.

extremely powerful and almost universally accepted rationale. Ergo, the widespread opinion that hereditarianism is irrelevant for politics is manifestly false.

Second, the knee-jerk opprobrium usually directed at any opposition to E is misplaced. Those who do not accept E (i.e., those who see nothing wrong in the existing social inequalities between blacks and whites) are occupying a demonstrably unacceptable moral position *only if* the truth of D is presupposed. Under the assumption that D is false, however, a new moral justification for E is necessary, and everything turns on whether it will be forthcoming or not. Ergo, the burden of proof is on the shoulders of the egalitarian.

Third, the egalitarians themselves ought to admit that their case would be weakened by the refutation of D. That is, they would have to agree that the moral urgency of eliminating inequalities is greater if D is true than if D is false. Suppose they deny this. This would mean that, in their own opinion, inequalities that are the result of discrimination are in no way more repulsive, morally, than those that are not the result of discrimination (say, those that are the result of different natural abilities). But this sounds absurd because it would imply that egalitarians would have to be totally indifferent to the question whether racial discrimination exists, for they would regard the alternative, the genetic source of inequalities, as *equally bad*.

Fourth, in commonsense morality the outrage at the existence of social inequalities is connected with the perceived *cause* of these inequalities. Discrimination on the basis of superficial, irrelevant characteristics is generally condemned. Inequalities produced in this way are denounced so resolutely and in such unison that it leaves nothing interesting for ethical discussion. But this consensus of outrage cannot be automatically transferred to social inequalities generated *in some other way*. Under the discrimination scenario, the proposal to abolish social inequalities wins by acclamation. If that scenario is false, however, a *reason* has to be given for abolishing social inequalities. Maybe such a reason exists, maybe it doesn't, but in any case we know in advance that, if it exists, it will have nothing to do with the reason that was behind our outrage at D-inequalities. This shows that the truth of hereditarianism would change the terms of the debate. In the new situation, those who want to continue fighting against social inequalities between groups could no longer simply count on being the champions of the noble cause. Now they would be forced to come up with a *new* argument to defend their position, and now they should expect to meet with many raised eyebrows rather than face the cheering audience as before.

Fifth, we are now in the position to see that Hume's law ("ought" cannot be derived from "is") does *not* mean, as it is often assumed, that "is" is irrelevant for "ought." Discoveries about facts *can* affect values. Science *can* have repercussions for morality and politics. The reason for this is that normative conclusions are usually not based on normative premises *alone* but on normative *and* descriptive premises. If one of the descriptive premises that was essential for grounding the normative conclusion is rejected, then the conclusion that followed from descriptive-cum-normative premises no longer follows. The justification for "ought" breaks down *because* of the change in "is."

Sixth, if there happens to be a correlation between a political (normative) position and the acceptance of a given factual belief, the explanation for this correlation is not necessarily that political interests are having an illegitimate influence on strictly scientific opinions. It may well be that the causal influence goes in the opposite direction, and that the factual belief (acquired bona fide) undermines one political position in the way explained above, and thereby supports the opposite political view.

Seventh, although it is true that after the factual part of the justification for a normative conclusion is discredited one can still stick to the normative conclusion by changing the normative part of the justification, this move may often be a sign of irrationality. Namely, if (a) your *only* reason for supporting the normative conclusion C was the conjunction of a normative premise N and a descriptive premise D, and (b) if even after you recognize that D is untenable you still support C, and (c) if your desire to continue supporting C is your exclusive reason for concocting a new justification, then your ground shifting is irrational (even if, luckily, you happen to stumble on a good justification). I suspect that such argumentative self-manipulation (what might be called "a conclusion in search of a premise") must be common when deeply held political beliefs are challenged. Whatever the way these beliefs are acquired in the first place, they hold a very central place in people's mentality, and by becoming part of their identity they are often not readily revisable. If ostensible reasons for these beliefs are undermined, what typically happens is that the subject immediately starts looking for ersatz grounds, and given the complexity of the matters, a "reason" that can minimally serve the purpose will usually be found.

Science and sensitivity

6.7 FROM GROUPS TO INDIVIDUALS

A standard reaction to the suggestion that there might be psychological differences between groups is to exclaim "So what?" Whatever these differences, whatever their origin, people should still be treated as individuals, and this is the end of the matter.

There are several problems with this reasoning. First of all, group membership is often a part of an individual's identity. Therefore, it may not be easy for individuals to accept the fact of a group difference if it does not reflect well on *their* group. Of course, whichever characteristic we take, there will usually be much overlap, the difference will be only statistical (between group averages), any group will have many individuals that outscore most members of other groups, yet individuals belonging to the lowest-scoring group may find it difficult to live with this fact. It is not likely that the situation will become tolerable even if it is shown that it is *not* product of social injustice. As Nathan Glazer said: "But how can a group accept an inferior place in society, even if good reasons for it are put forth? It cannot" (Glazer 1994: 16). In addition, to the extent that the difference turns out to be heritable there will be more reason to think that it will not go away so easily (see chapter 5). It will not be readily eliminable through social engineering. It will be modifiable in principle, but not locally modifiable (see section 5.3 for the explanation of these terms). All this could make it even more difficult to accept it.

Next, the statement that people should be treated as individuals is certainly a useful reminder that in many contexts direct knowledge about a particular person eclipses the informativeness of any additional statistical data, and often makes the collection of this kind of data pointless. The statement is fine as far as it goes, but it should not be pushed too far. If it is understood as saying that it is a *fallacy* to use the information about an individual's group membership to infer something about that individual, the statement is simply wrong. Exactly the opposite is true: it is a fallacy *not* to take this information into account.

Suppose we are interested in whether John has characteristic F. Evidence E (directly relevant for the question at hand) indicates that the probability of John having F is p. But suppose we also happen to know that John is a member of group G. Now elementary probability theory tells us that if we want to get the *best* estimate of the probability that John has F we *have* to bring the group information to bear on the issue. In calculating the desired probability we have to take into account (a) that

John is a member of G, *and* (b) what proportion of G has F. Neglecting these two pieces of information would mean discarding potentially relevant information. (It would amount to violating what Carnap called "the requirement of total evidence.") It may well happen that in the light of this additional information we would be forced to revise our estimate of probability from p to p*. Disregarding group membership is at the core of the so-called "base rate fallacy," which I will describe using Tversky and Kahneman's taxicab scenario (Tversky & Kahneman 1980).

In a small city, in which there are 90 green taxis and 10 blue taxis, there was a hit-and-run accident involving a taxi. There is also an eyewitness who told the police that the taxi was blue. The witness's reliability is 0.8, which means that, when he was tested for his ability to recognize the color of the car under the circumstances similar to those at the accident scene, his statements were correct 80 percent of the time. To reduce verbiage, let me introduce some abbreviations: B = the taxi was blue; G = the taxi was green; W_B = witness said that the taxi was blue.

What we know about the whole situation is the following:

(1) $p(B) = 0.1$ (the *prior* probability of B, *before* the witness's statement is taken into account)
(2) $p(G) = 0.9$ (the prior probability of G)
(3) $p(W_B/B) = 0.8$ (the reliability of the witness, or the probability of W_B, given B)
(4) $p(W_B/G) = 0.2$ (the probability of W_B, given G)

Now, given all this information, what is the probability that the taxi was blue in that particular situation? Basically we want to find $p(B/W_B)$, the *posterior* probability of B, i.e., the probability of B *after* W_B is taken into account. People often conclude, wrongly, that this probability is 0.8. They fail to take into consideration that the proportion of blue taxis is pretty low (10 percent), and that the true probability must reflect that fact.[29] A simple rule of elementary probability, Bayes' theorem, gives the formula to be applied here:

$$p(B/W_B) = p(B) \times p(W_B/B) / [p(B) \times p(W_B/B) + p(G) \times p(W_B/G)].$$

Therefore, the correct value for $p(B/W_B)$ is 0.31, which shows that the usual guess (0.8 or close to it) is wide of the mark.

---

[29] For an argument that the base rate fallacy is much less frequent than many psychologists have thought, see Koehler 1996.

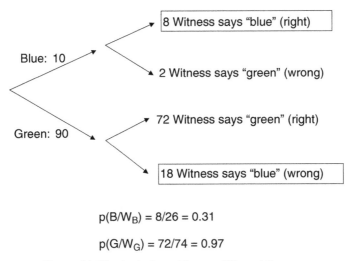

$p(B/W_B) = 8/26 = 0.31$

$p(G/W_G) = 72/74 = 0.97$

Figure 6.1 The taxicab problem and Bayes' theorem.

It is easier to understand that 0.31 is the correct answer by looking at Figure 6.1. Imagine that the situation with the accident and the witness repeats itself 100 times. Obviously, we can expect that the taxi involved in the accident will be blue in 10 cases (10 percent), while in the remaining 90 cases it will be green. Now consider these two different kinds of cases separately. In the top section (blue taxis), the witness recognizes the true color of the car 80 percent of the times, which means in 8 out of 10 cases. In the bottom section (green taxis), he again recognizes the true color of the car 80 percent of the times, which here means in 72 out of 90 cases. Now count all those cases where the witness declares that the taxi is blue, and see how often he is right about it. Then simply divide the number of times he is right when he says "blue" with the overall number of times he says "blue," and this will immediately give you $p(B/W_B)$. The witness gives the answer "blue" 8 times in the upper section (when the taxi is indeed blue), and 18 times in the bottom section (when the taxi is actually green). Therefore, our probability is: $8/(8 + 18) = 0.31$.

It may all seem puzzling. How can it be that the witness says the taxi is blue, his reliability as a witness is 0.8, and yet the probability that the taxi is blue is only 0.31? Actually there is nothing wrong with the reasoning. It is the lower prior frequency of blue taxis that brings down the probability of the taxi being blue, and that is that. Bayes' theorem is a mathematical truth. Its application in this kind of situation is beyond dispute. Any remaining doubt will be dispelled by inspecting Figure 6.1 and seeing that if you trust the witness when he says "blue" you will indeed be more often wrong than right. But notice that you have excellent reasons to trust the witness if he says "green" because in that case he will be right 97 percent of the time! It all follows from the difference in prior probabilities for "blue" and "green." There is a consensus that neglecting prior probabilities (or base rates) is a logical fallacy.

But if neglecting prior probabilities is a fallacy in the taxicab example, then it cannot stop being a fallacy in other contexts. Oddly enough, many people's judgment actually changes with context, particularly when it comes to inferences involving social groups. The same move of neglecting base rates that was previously condemned as the violation of elementary probability rules is now praised as reasonable, whereas applying the Bayes' theorem (previously recommended) is now criticized as a sign of irrationality, prejudice and bigotry.

A good example is racial or ethnic profiling,[30] the practice that is almost universally denounced as ill advised, silly, and serving no useful purpose. This is surprising because the inference underlying this practice has *the same logical structure* as the taxicab situation. Let me try to show this by representing it in the same format as Figure 6.1. But first I will present an example with some imagined data to prepare the ground for the probability question and for the discussion of group profiling.

Suppose that there is a suspicious characteristic E such that 2 percent terrorists (T) have E but only 0.002 percent non-terrorists $(-T)$ have E. This already gives us two probabilities: $p(E/T) = 0.02$; $p(E/-T) = 0.00002$. How useful is E for recognizing terrorists? How likely is it that someone is T if he has E? What is $p(T/E)$? Bayes' theorem tells us that the answer depends on the percentage of terrorists in a population. (Clearly, if everybody is a terrorist, then $p(T/E) = 1$; if no one is a terrorist, then $p(T/E) = 0$; if some people are T and some $-T$, then $1 > p(T/E) > 0$.) To activate the group question, suppose that there are two groups, A and B, that have different percentages of terrorists (1 in 100, and 1 in 10,000, respectively). This translates into different probabilities of an arbitrary member of a group being a terrorist. In group A, $p(T) = 0.01$ but in group B, $p(T) = 0.0001$. Now for the central question: what will $p(T/E)$ be in A and in B? Figures 6.2a and 6.2b provide the answer.

---

[30] What is ethnic profiling? It is a situation where different ethnic groups have different proportions of people with a given characteristic, and where in estimating the probability that a particular individual has that characteristic, information about his ethnicity is factored in.

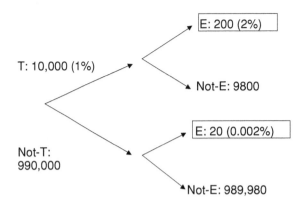

p(T/E) = 200/220 = 0.91

Figure 6.2a A terrorist suspect (Group A).

In group A, the probability of a person with characteristic E being a terrorist is 0.91. In group B, this probability is 0.09 (more than ten times lower). The group membership matters, and it matters a lot.

Test your intuitions with a thought experiment: in an airport, you see a person belonging to group A and another person from group B. Both have suspicious trait E but they go in opposite directions. Whom will you follow and perhaps report to the police? Will you (a) go by probabilities and focus on A (committing the sin of racial or ethnic profiling), or (b)

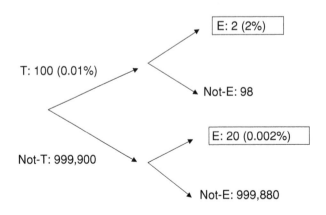

p(T/E) = 2/22 = 0.09

Figure 6.2b A terrorist suspect (Group B).

follow political correctness and flip a coin (and feel good about it)? It would be wrong to protest here and refuse to focus on A by pointing out that most As are *not* terrorists. This is true but irrelevant. Most As that have E *are* terrorists (91 percent of them, to be precise), and this is what counts. Compare that with the other group, where out of all Bs that have E, less than 10 percent are terrorists.

To recapitulate, since the two situations (the taxicab example and the social groups example) are similar in all relevant aspects, consistency requires the same answer. But the resolute answer is already given in the first situation. All competent people speak with one voice here, and agree that in this kind of situation the witness's statement is only part of the relevant evidence. The proportion of blue cars *must* also be taken into account to get the correct probability that the taxi involved in the accident was blue. Therefore, there is no choice but to draw the corresponding conclusion in the second case. E is only part of the relevant evidence. The proportion of terrorists in group A (or B) *must* also be taken into account to get the correct probability that an individual from group A (or B) is a terrorist.

The "must" here is a conditional "must," not a categorical imperative. That is, you must take into account prior probabilities *if* you want to know the true posterior probability. But sometimes there may be other considerations, besides the aim to know the true probability. For instance, it may be thought unfair or morally unacceptable to treat members of group A differently from members of group B. After all, As belong to their ethnic group without any decision on their part, and it could be argued that it is unjust to treat *every* A as more suspect just because a very small proportion of terrorists among As happens to be higher than an even lower proportion of terrorists among Bs. Why should some people be inconvenienced and treated worse than others only because they share a group characteristic, which they did not choose, which they cannot change, and which is in itself morally irrelevant?

I recognize the force of this question. It pulls in the opposite direction from Bayes' theorem, urging us *not* to take into account prior probabilities. The question which of the two reasons (the Bayesian or the moral one) should prevail is very complex, and there is no doubt that the answer varies widely, depending on the specific circumstances and also on the answerer. I will not enter that debate at all because it would take us too far away from our subject.

The point to remember is that when many people say that "an individual can't be judged by his group mean" (Gould 1977: 247), that "as individuals we are all unique and population statistics do not apply" (Venter 2000), that "a person should not be judged as a member of a group but as an individual" (Herrnstein & Murray 1994: 550), these statements sound nice and are likely to be well received but they conflict with the hard fact that a group membership sometimes does matter. If scholars wear their scientific hats when denying or disregarding this fact, I am afraid that rather than convincing the public they will more probably damage the credibility of science.

It is of course an empirical question *how often* and *how much* the group information is relevant for judgments about individuals in particular situations, but before we address this complicated issue in specific cases, we should first get rid of the wrong but popular idea that taking group membership into consideration (when thinking about individuals) is *in itself* irrational or morally condemnable, or both. On the contrary, in certain decisions about individuals, people "would have to be either saints or idiots not to be influenced by the collective statistics" (Genovese 1995: 333).

Before we move on, though, just a reminder of how we got involved in the topic in the first place (as far as we did). Discussions about heritability reach their most sensitive point, politically speaking, with the issues about group differences. According to an influential opinion, the nervousness about political implications of group differences (and their possibly high heritability) is entirely unjustified for the simple reason that people should always and exclusively be judged *as individuals*. It is argued that any consideration of group characteristics (like ethnic group membership) when dealing with individuals springs from irrational prejudice and cognitive defect. My point is that all is not that simple. Whether we like it or not, group facts sometimes do affect probabilities of individuals possessing socially significant characteristics. Of course, if we knew *everything* about a particular individual, then the information about groups to which that individual belongs would fade into irrelevance. An omniscient god would have no use for Bayes' theorem. We mortals, however, often have to deal with people about whom we know relatively little, and in these situations relying on prior probabilities from group data is *epistemically reasonable*.

My example with terrorism illustrates neatly the application of Bayes' theorem to social groups, but it is not the best example for a context where heritability might come into play. For this purpose, a better illustration would be groups with different rates of violent crime, where the difference might involve at least a partial genetic explanation. Interestingly, in cases where groups are recognizably different in some genetic respects and where the issue of hetitability logically arises, our response is not always consistent. For instance, the biological explanation for the higher incidence of violent crime among men than among women (and the use of these data for probability inferences) encounters little ideological resistance, but the corresponding (and structurally similar) approach in the case of racial groups is considered not just morally unacceptable but epistemically defective as well. (To see how group characteristics influence the relevant probabilities, see a nice simulation of these dependencies that is developed by my brother and my computer *consigliere*, Ante Sesardic, at www.ln.edu.hk/philosophy/staff/sesardic/bayes. It should not be assumed, however, that Ante would necessarily agree with all the views I express in this section.)

In sum, the interest in acquiring knowledge can clash with the interest in treating all people fairly and equally. But this conflict of theoretical and practical reason cannot be resolved by switching off the former or by pretending that it does not exist. True, sometimes the statistical difference between groups will be so small that it will not significantly affect probabilities in which we may be interested. In such situations the information about groups will be all but useless, but even then we would first have to collect and review the data – in order to be able to conclude that they are useless. And if a between-group difference happens to be non-negligible, the discovery that it is highly heritable would tend to exacerbate the conflict between the cognitive and ethical demands. For in the case of a genetic explanation, the group-based prior probabilities which produce such moral outrage and which tend to be suppressed, swept under the carpet, or irrationally condemned as irrelevant, would prove to be even more persistent and less readily modifiable by environmental manipulation.

6.8 FAIR, THEREFORE BIASED?

Some people's moral sensitivities are offended by yet another group–individual inference that is licensed by standard probability theory. Figuratively speaking, this inference (known as "regression to the mean") violates John Rawls's Difference Principle because it urges us to revise our judgments about individuals in a way that "harms" the worst-off group.

What is regression to the mean? It tells us that if there are two different populations with different trait means and if the trait is normally distributed in both populations, then in any unbiased measurement that is not 100 percent reliable, measured individual scores should always be "corrected" by taking into account the relevant mean. The scores should be regressed to the mean of the population to which the individual in question belongs. The procedure is statistically uncontroversial and gives the best estimate of the "true score" on the basis of measurement.

Expressed in general terms, the inference looks benign enough, but the feeling may easily change when regression to the mean is applied to the highly explosive context of the racial gap in IQ. It is an accepted empirical fact that American blacks, as a group, have approximately one standard deviation lower IQ than whites. The black mean is 85 and the white mean is 100. Now regression to the mean tells us here that if there is a black person and a white person with the *same* measured IQ, then if we know nothing else about these two persons and if we want to get the best estimates of their true IQs on the basis of their measured scores, we should ascribe a *lower* IQ to the black person than to the white person. This is a purely mathematical consequence: if there is measurement error (i.e., the test is not perfectly reliable), the real IQ will more probably lie somewhere toward the mean than away from the mean.[31] Since the black mean is lower than the white mean, black true scores will have a "downward pull" compared to white true scores.

The amount of regression depends on two things: first, the test reliability (r), and second, the difference between the measured score (M) and the mean (μ). The amount of regression is given by the formula: $(1 - r)(M - \mu)$. So, the regression will be larger, the less reliable the test, and the larger the difference between the measured score and the mean. The amount of regression decreases with test reliability and is directly proportional to the difference between the measured score and the group mean. Since the black mean is lower than the white mean, correction will be "less favorable" for blacks than for whites. That is, the score of everyone whose IQ is below the average of his group[32] will be slightly corrected upward and the score of everyone whose IQ is above the average of his group will be slightly corrected downward, but on average the upward correction will be greater for whites and the downward correction will be greater for blacks.

[31] Think about it this way. Take a person with the measured IQ of 140. Suppose you know that the measurement error is 30 IQ points but you don't know in which direction. Isn't it obvious that it is much more likely that the person's IQ is 110 (very common) than 170 (extremely rare)?

[32] Every individual belongs to many groups, of course, and if we want to estimate his most probable IQ, the question arises about how to choose the most appropriate reference group for this purpose. I will neglect these problems, however, because I am here just interested in how the information about *race* affects relevant probabilities in this context.

This may seem very unfair. It looks as if those who are already unfortunate enough to belong to the worst-off group are pushed even further down by some kind of statistical legerdemain. Doesn't it amount to Robin Hoodery in reverse, robbing the poor in order to compound the existing inequalities? Apparently, it would be nice if there were some way to criticize the mathematical inference and show that the "inequitable" derivation of probabilities is fallacious.

A valiant effort in that direction was undertaken by Ned Block and Gerald Dworkin. They tried to prove, against the consensus of statisticians, that even under the assumption that IQ tests are *not* biased against blacks (i.e., that "tests measure intelligence equally well in individuals irrespective of color"), the right statistical procedure would be to have the lower average IQ score of blacks, as a group, corrected *upward*. The conclusion may be pleasing in some sense, but it rings false. How can it be, if there is no racial bias, that there is a need for a correction on the basis of race? Block and Dworkin found this conclusion "rather surprising" themselves (Block & Dworkin 1976b: 458), but this did not stop them from defending it with a great deal of confidence.

They start with the truth that, on average, people with below-average IQs have their intelligence (their "true score") underestimated by IQ tests. This is regression to the mean, pure and simple. They further say, again correctly, that if the reliability of the test is .5 (what they call the correlation between IQ and intelligence), then any person with a measured score of 85 and belonging to a population with the IQ mean of 100 should be inferred to have the "true intelligence" of 92.5. Very well. But then comes the crucial part of their argument:

> the expected intelligence of blacks is no less than that of a group of persons chosen at random which happens to have an average IQ of 85. It follows that the black intelligence expected on the basis of IQ is higher than 85. If the IQ–intelligence correlation is .5, the black–white intelligence gap expected on the basis of IQ is half the black–white gap ... The application of this point [regression to the mean] to color bias is that *blacks do count as a randomly chosen IQ 85 group for our purposes because there is no reason to think their intelligence is lower than the IQ 85 population as a whole.* (Block & Dworkin 1976b: 459, 529 [note] – italics added)

After making this point Block and Dworkin wonder why, despite its "obvious importance," they have never seen their point made in psychometric literature. The explanation is "not hard to find," they say: it is operationalism and instrumentalism prevalent among psychometricians. But maybe there is an alternative explanation. Could it be that the reason why they have never seen their point made is simply that it is wrong, and blatantly so? Let us explore this possibility.

In regression to the mean, what regresses is individual scores, *not* the mean. It is regression *to* the mean (or regression *toward* the mean), not regression *of* the mean (or regression *away from* the mean). So the idea that, despite the absence of bias, the mean of a given population (blacks) should be corrected upward is fundamentally wrong. If a given population has a mean IQ of 85, then the best estimate of the average true intelligence in that population is the mean itself (85). How did Block and Dworkin manage to go astray so badly?

The mistake is committed in the italicized part: "*blacks do count as a randomly chosen IQ 85 group for our purposes because there is no reason to think their intelligence is lower than the IQ 85 population as a whole.*" If a group of people with an IQ of 85 is actually taken out of the population with the mean IQ of 100, then indeed any individual IQ score in that group should be corrected upward. But this reasoning clearly does not apply to American blacks because they do *not* belong to the population with the mean IQ of 100. Their mean IQ is 85. Therefore any 85 score in this group is identical to the mean, and obviously there can be no regression to the mean if the score is at the mean already.

Moreover, as Jensen pointed out (1980: 422–423), Block and Dworkin did not understand that the unreliability of the test would actually lead to the *underestimation* of the racial IQ difference, not overestimation. This is simply a consequence of the statistical fact that unreliable tests cannot capture the measured effect in its entirety. For this reason, if one wants to correct for unreliability, the *true* inter-racial difference should be inferred to be greater than the *observed* one, the correction factor increasing in proportion to the test's unreliability. What Block and Dworkin did in that situation was to use an entirely improper calculation, according to which the difference in true score means turns out to be *smaller* than the difference between observed score means, *instead of being greater* (see, e.g., Cohen 1988: 537). They started by assuming that IQ tests are not biased against blacks and that the reliability of IQ tests is .5, and then in the attempt to infer the true size of the white–black difference from the observed difference (15 points), they applied the wrong formula. In mathematical terms, they obtained the "true" difference by *multiplying* the observed racial IQ difference of 15 points by .5, *instead of dividing* it by .5, and because of that mistake they inferred that the true racial gap was 7.5 points,[33] whereas the statistically correct answer would be 30 points (their estimate thus being wrong by a factor of 4).

It is ironic that the two philosophers who entered the IQ debate with the intention of clearing up conceptual confusions managed only to muddle the issues further with egregiously fallacious reasoning of their own.

---

[33] Actually the reliability of IQ tests is much higher than .5 (the value chosen by Block and Dworkin), so the difference between black and white "true" scores would not really be much larger than the difference between their observed scores.

7

Conclusion

A great deal of intelligence can be invested in ignorance when the need for illusion is deep.

Saul Bellow

At the end of this book many a reader may feel that my approach should have been more balanced. Even if I am right in pointing out many weaknesses of environmentalist criticisms of heritability, doesn't fair scholarship require that problematic aspects of hereditarianism be addressed as well? Surely, bad arguments cannot be a "privilege" of one side in the debate.

In my defense, let me remind you that my goal was *not* to offer a comprehensive discussion of the nature–nurture problem. I focused just on a small segment of that controversy. As a philosopher of science, I found it interesting to scrutinize very general *methodological* arguments that are often used to short-circuit the debate in the attempt to undermine one of the rival positions, without going into empirical details at all. And precisely here is the source of the disparity. It is *only* environmentalists who want to use this kind of methodological shortcut. Hereditarians are quite happy to let the empirical evidence decide the matter. So the imbalance of my approach is the result of an existing asymmetry, not of my partiality.

Whereas methodological arguments purported to prove that heritability claims are meaningless, confusing, or causally uninterpretable, a completely opposite criticism is that they are trivial. Some people recently started to claim that it has always been clear that both genes and environments affect phenotypes, and that heritabilities substantially greater than zero and substantially smaller than one were exactly what reasonable people always expected:

Paradoxically, this is a corner of science where the "expert" has usually been more wrong than the layman. Ordinary people have always known that education matters, but equally they have always believed in some innate ability. It is the experts who have taken extreme and absurd positions at either end of the spectrum. (Ridley 2000: 80)

Sensible people reject both the hereditarian claim that genes explain everything and the environmentalist claim that they explain nothing – they stand for a reasonable middle ground between these absurdly simplistic extremes. (Paul 1998: 82)

The postulated symmetry between "absurd" positions is a historical myth. The extreme position actually existed only at one end of the spectrum: the environmentalist end. Many experts and ordinary people have believed that human psychological differences are *exclusively* the result of differences in environmental influences. On the other hand, no scholar has ever claimed that *all* psychological differences are caused by genetic differences. Hans Eysenck was right when he protested against this kind of distortion:

The assumption that there were 100% hereditarians among the behavioural geneticists will not stand up. Having been in the middle of that battle, and having known many of the leading participants, I cannot think of one who would have assumed anything so obviously silly as a 100% genetic determination of any behavioural variable. The imputation that there were any 100% hereditarians is simply untrue; there were, and are, 100% environmentalists, and they carried the day for many years, largely ignoring the available evidence, or misrepresenting it. (Eysenck 1994: 582)

Do we here have two equally absurd extremes, asserting immoderate and egregious nonsense? Not quite. Like the ski resort full of girls hunting for husbands and husbands hunting for girls, the situation is not as symmetrical as it might seem. No psychologist, geneticist, or biologist has ever asserted that nature was all, and nurture played no part, even Galton's somewhat extreme statement merely asserts that nature is more important than nurture. (Eysenck 1998: 29–30)

In support of Eysenck's diagnosis recall that the title of one of Galton's main works was *English Men of Science: Their Nature **and Nurture***.

Radical environmentalists not only existed, they completely dominated the academic scene, in particular after World War II. James Watson says that in the war's aftermath "even thinking about human behavior as having a genetic basis was a no-no" (Watson 2003: xxi). Irving Gottesman recalls that his first efforts in behavior genetics "were rejected by editors as anachronistic and as an attempt to resurrect the defunct nature versus nurture battles of the 1920s and 1930s" (Gottesman 2002: xvi). Sandra Scarr writes:

> My interest in the possibility of genetic behavioral differences began when, as an undergraduate, I was told that there were none. The social science view of the time was that genetics might set limits on species but that each individual within the species was endowed with everything that was important to develop into a beggar, king, attorney, or con artist. (Scarr 1987: 221)

It is easy to forget how until quite recently the research into heritability of psychological traits encountered acrimonious resistance. The situation was so bad that in 1972 a group of very distinguished scientists (which included, among others, Francis Crick, Otis Dudley Duncan, Jacques Monod, and Paul Meehl) issued a warning about "suppression, censure, punishment and defamation [that] are being applied against scientists who emphasize the role of heredity in human behavior" (Page 1972: 660). They said: "it is virtually a heresy to express a hereditarian view, or to recommend a further study of the biological bases of behavior. A kind of orthodox environmentalism dominates the liberal academy, and strongly inhibits teachers, researchers and scholars from turning to biological explanations or efforts."

Ironically, it is the very success of behavior genetics in documenting the pervasive influence of genes on human psychology that makes us today develop a "false memory syndrome" and think that we must have known this all along. Eric Turkheimer (Turkheimer 1998: 786; 2000: 160) proclaims as "the first law of behavior genetics" that all human behavioral traits are heritable. Indeed, nowadays a discovery that a trait is heritable is no longer newsworthy. It is a result that is expected, unsurprising, we could even say boring. But it was not always like that. There was no necessity that things would turn this way. There was nothing "natural" or "self-evident" about genetic influences being so important for psychological differences. For all that matters, the empirical evidence could well have gone in the opposite direction. But it didn't. Moreover, everyone expected a different result initially. The "astounding fact" of massive heritability (Turkheimer 1998: 785) took by surprise not only opponents of behavior genetics but behavior geneticists themselves as well.

The empirical support for the massive presence of genetic influences turned out to be so powerful, consistent, and unequivocal that in hindsight the conclusion looks to some people acceptable to the point of obviousness or even triviality, and it all somehow detracted from the success of behavior genetics. The research on heritability is, to use Nietzsche's phrase, like "a wrestler who turned his own strength against himself and was checked and wounded by his own victory" (Nietzsche 1973: 203).

To see how much the situation changed, just remember that it was only two decades ago that in a highly influential book (Rose et al. 1984: 116) it was claimed that, for all we know, the heritability of IQ may well be zero. The authors of the book were praised by a prominent philosopher of biology for their "brilliant" and "lucid" arguments (Kitcher 1984: 9) as well as by a distinguished scientist who said that their general view was "correct," "shared by many others," and "rather moderate" (Bateson 1985: 59). Closer to the present moment, the idea that human cognitive ability is substantially heritable was still disputed in 1996 by the National Institutes of Health–Department of Energy Joint Working Group on the Ethical, Legal, and Social Implications of Human Genome Research (ELSI Working Group). Their statement[1] to that effect was published in the journal *Science* (Nelkin & Andrews 1996: 13). Most critics today would not go that far. They usually accept that heritability of many psychological traits is substantial but they argue that heritability has no important implications about genetic causation, between-group heritability, limited malleability, etc. However, the fact that, as late as the mid-nineties, a group of prominent scientists (geneticists working on the Human Genome Project) could bring themselves to publicly express doubts about heritability of IQ shows best that the claim is far from being trivial or self-evident.

We should also recall that for a long time the only environmental influences that were thought to be an important alternative to genetic causes were things like differences in socioeconomic status, schools, parental style, home characteristics, etc. They are now called "shared environments" (see chapter 5), because they are shared by children living in the same family. No one ever suspected that environmental differences within the family ("non-shared environments") could have a substantial impact on, say, IQ or personality differences. For this reason, heritability was sometimes estimated by simply measuring the shared environmental influences and then taking the remainder to be due to genes. So, heritability was either obtained directly (from the correlation of MZ twins reared

[1] See the correction of authorship in *Science*, February 16, 1996.

apart) or indirectly (by subtracting from 1 the correlation of genetically unrelated children reared in the same home). Christopher Jencks and Mary Jo Bane noticed that these two methods did not produce the same result, and here was their advice how to reconcile the two estimates: "The most reasonable assumption is that the true effect of heredity is somewhat less than that suggested by twin studies but somewhat more than that suggested by studies of unrelated children in the same home" (Jencks & Bane 1976: 331).

And now comes the interesting thing. If we follow their advice when dealing with the *current* data obtained with these two methods, we would actually have to conclude that the heritability of IQ is somewhat lower than 75 percent (the direct measure, from twin studies) but somewhat higher than 100 percent (the indirect measure, from adoption studies)! The absurd heritability estimate that exceeds 100 percent is due to the fact that the influence of shared environments on IQ differences in adulthood is essentially zero. Hence, heritability comes out maximal if calculated by subtraction $(1 - 0)$, and even higher if corrected upward (in accordance with Jencks and Bane's suggestion).

Christopher Jencks was one of the most sophisticated participants in the debate on IQ in the seventies, and at that time his proposal about how to get the optimal estimate of heritability was not so unreasonable at all. That it leads to incoherence now is the best testimony that we have indeed come a long way since then.

The research on heritability has already produced many surprising discoveries, raised new issues with possibly far-reaching implications, and stimulated novel ways of thinking even in the areas of psychology that have nothing to do with genetics. For example, Judith Harris's book *The Nurture Assumption* (Harris 1998) – which challenged the orthodoxy in child psychology and which was called "a paradigm shifter" and was predicted to be "a turning point in the history of psychology" – was actually, in the author's own words, her "attempt to solve a puzzle turned up by behavioral geneticists" (Harris 1996). All these fascinating developments coming out of heritability research are sometimes hidden from view by the tall weeds of methodological pseudo-criticism eagerly cultivated by a number of scientists, philosophers of science, and public intellectuals.

I hope that with some of these bad plants weeded out it will be easier to approach the controversial issues with a more open mind. It will be easier to recognize that there is no royal road to the truth about heritability and its implications, no instant refutation that would by-pass empirical investigations. Hereditarianism has too long been dismissed out of hand because of its alleged methodological defects. The arguments used for that purpose were accepted too readily, without sufficient critical reflection. These arguments sounded "intuitively right" as well as politically opportune, and somehow it didn't matter at all that they were based on fallacious reasoning, distortions of interpretation, and ignorance of the relevant literature. It is particularly disappointing that this consensus of the uninformed persisted for such a long time in philosophy of science, the field that is supposed to manifest a high level of conceptual sophistication and logical rigor.

References

Alland, A. 2002, *Race in Mind: Race, IQ, and Other Racisms*, New York, Palgrave Macmillan.

Allen, G. E. 1990, "Genetic Indexing of Race Groups is Irresponsible and Unscientific," *Scientist*, May 14.

1994, "Social Origins of Genetic Determinism," in E. Tobach and B. Rosoff (eds.), *Challenging Racism and Sexism: Alternatives to Genetic Explanations*, New York, The Feminist Press at CUNY.

Alper, J. S. 1998, "Genes, Free Will, and Criminal Responsibility," *Social Science & Medicine* 46: 1599–1611.

Anastasi, A. 1958, "Heredity, Environment, and the Question 'How?'" *Psychological Review* 65: 197–208.

Antony, L. 1997, Review of *Measured Lies*, *Personnel Psychology* 50: 481–485.

2000, "Natures and Norms," *Ethics* 11: 8–36.

Appiah, K. A. 1995, "Straightening out *The Bell Curve*," in R. Jacoby and N. Glauberman (eds.), *The Bell Curve Debate*, New York, Random House.

1996, "Race, Culture, Identity: Misunderstood Connections," in *The Tanner Lectures on Human Values*, vol. 17, Salt Lake City, University of Utah Press.

Ariew, A. 1996, "Innateness and Canalization," *Philosophy of Science (Proceedings)* 63: S19–S27.

Baker, C. 2004, *Behavioral Genetics: An Introduction to How Genes and Environments Interact through Development to Shape Differences in Mood, Personality, and Intelligence*, Washington, D.C., AAAS.

Barbujani, G., Magagni, A., Mirch, E., and Cavalli-Sforza, L. L. 1997, "An Apportionment of Human DNA Diversity," *Proceedings of the National Academy of Sciences* 94: 4516–4519.

Bateson, P. 1985, "Sociobiology: The Debate Continues," *New Scientist*, January 24.

2001a, "Behavioral Development and Darwinian Evolution," in S. Oyama, P. E. Griffiths, and R. D. Gray (eds.), *Cycles of Contingency: Developmental Systems and Evolution*, Cambridge, Mass., MIT Press.

2001b, "Where Does our Behavior Come from?" *Journal of Bioscience* 26: 561–570.

2002, "The Corpse of a Wearisome Debate," *Science* 297: 2212–2213.

Bateson, P. P. G. and Martin, P. R. 2000, *Design for a Life: How Behavior and Personality Develop*, New York, Simon & Schuster.

Baumeister, A. A. and Bacharach, V. R. 2000, "Early Generic Educational Intervention Has No Enduring Effect on Intelligence and Does Not Prevent Mental Retardation: The Infant Health and Development Program," *Intelligence* 28: 161–192.

Beckwith, J. 1993, "A Historical View of Social Responsibility in Genetics," *Bioscience* 43: 327–333.

Beckwith, J. and Alper, J. S. 2002, "Genetics of Human Personality: Social and Ethical Implications," in J. Benjamin, R. P. Ebstein, and R. H. Belmaker (eds.), *Molecular Genetics and Human Personality*, Washington, D.C., American Psychiatric Publishing, Inc.

Bell, A. E. 1977, "Heritability in Retrospect," *Journal of Heredity* 68: 297– 300.

Bennett, J. H. 1983, *Natural Selection, Heredity, and Eugenics*, Oxford, Clarendon Press.

Berkeley, G. 1996, *Principles of Human Knowledge and Three Dialogues*, Oxford, Oxford University Press.

Berkowitz, A. 1996, "Our Genes, Ourselves," *Bioscience* 46: 42–51.

Billings, P. R., Beckwith, J., and Alper, J. S. 1992, "The Genetic Analysis of Human Behavior: A New Era?" *Social Science and Medicine* 35: 227–238.

Blackburn, S. 1996, *The Oxford Dictionary of Philosophy*, Oxford, Oxford University Press.

2002, "Meet the Flintstones," *New Republic*, November 25.

Blinkhorn, S. 1982, "What Skullduggery?" *Nature* 296: 506.

Block, N. J. 1974, Letter, *Psychology Today*, March.

1995, "How Heritability Misleads About Race," *Cognition* 56: 99–128.

Block, N. J. and Dworkin, G. (eds.) 1976a, *The IQ Controversy: Critical Readings*, New York, Pantheon.

1976b, "IQ, Heritability and Inequality," in N. J. Block and G. Dworkin (eds.), *The IQ Controversy*, New York, Pantheon.

Boas, F. 1912, "Changes in the Bodily Form of Descendants of Immigrants," *American Anthropologist* 14: 530–562.

Bodmer, W. F. and Cavalli-Sforza, L. L. 1970, "Intelligence and Race," *Scientific American* 223: 19–29.

Boer, S., Farrell, D., Garner, R., et al. 1990, Letter to the editor, *Proceedings and Addresses of the American Philosophical Association*, June.

Boomsma, D. I. and Martin, N. G. 2002, "Gene–Environment Interactions," in H. D'haenen, J. A. den Boer, and P. Willner (eds.), *Biological Psychiatry*, New York, Wiley.

Bouchard, T. J. 1987, "The Hereditarian Research Program: Triumphs and Tribulations," in S. Modgil and C. Modgil (eds.), *Arthur Jensen: Consensus and Controversy*, New York, Falmer Press.

1997, "IQ Similarity in Twins Reared Apart: Findings and Responses to Critics," in R. J. Sternberg and E. Grigorenko (eds.), *Intelligence, Heredity and Environment*, Cambridge, Cambridge University Press.

1998, "Genetic and Environmental Influences on Adult Intelligence and Special Mental Abilities," *Human Biology* 70: 257–279.

Bouchard, T. J. and Loehlin, J. C. 2001, "Genes, Evolution, and Personality," *Behavior Genetics* 31: 243–273.

Bowler, P. J. 1989, *The Mendelian Revolution: The Emergence of Hereditarian Concepts in Modern Science and Society*, Baltimore, Md., Johns Hopkins University Press.

Brace, C. L. 2001, Review of Joseph L. Graves, *The Emperor's New Clothes: Biological Theories of Race at the Millennium, American Scientist* 89: 277.

Brand, C. 1996, *The g Factor*, New York, Wiley.

Brandon, R. 1996, *Concepts and Methods in Evolutionary Biology*, Cambridge, Cambridge University Press.

Brockman, J. (ed.) 1995, *The Third Culture*, New York, Simon & Schuster.

Brody, N. 1992, *Intelligence*, San Diego, Academic Press.

1997, "Malleability and Change in Intelligence," in Nyborg, H. (ed.), *The Scientific Study of Human Nature: Tribute to Hans J. Eysenck at Eighty*, New York, Pergamon.

Brody, N. and Crowley, M. J. 1995, "Environmental (and Genetic) Influences on Personality and Intelligence," in D. H. Saklofske and M. Zeidner (eds.), *International Handbook of Personality and Intelligence*, New York, Plenum.

Brown, J. R. 1998, "Intellectual Tithing," *International Studies in the Philosophy of Science* 12: 5.

2001, *Who Rules in Science? An Opinionated Guide to the Wars*, Cambridge, Mass., Harvard University Press.

Bruer, J. T. 1999, *The Myth of the First Three Years: A New Understanding of Early Brain Development and Lifelong Learning*, New York, Free Press.

Buchanan, A., Brock, D. W., Daniels, N., and Winkler, D. 2002, *From Chance to Choice: Genetics and Justice*, Cambridge, Cambridge University Press.

Bunge, M. 1996, "In Praise of Intolerance to Charlatanism in Academia," in P. R. Gross, N. Levitt, and M. W. Lewis (eds.), *The Flight from Science and Reason*, Baltimore, Md., Johns Hopkins University Press.

Burian, R. M. 1981, "Human Sociobiology and Genetic Determinism," *Philosophical Forum* 13: 43–66.

Burks, B. S. 1938, "On the Relative Contributions of Nature and Nurture to Average Group Differences in Intelligence," *Proceedings of the National Academy of Sciences* 24: 276–282.

Burks, B. S. and Kelley, T. L. 1928, "Statistical Hazards in Nature–Nurture Investigations," *Twenty-Seventh Yearbook of the National Society for the Study of Education*, Bloomington, University of Indiana Press.

Burks, B. S. and Tolman, R. 1932, "Is Mental Resemblance Related to Physical Resemblance in Sibling Pairs?" *Journal of Genetic Psychology* 40: 3–15.

Callebaut, W. (ed.) 1993, *Taking the Naturalist Turn*, Chicago, University of Chicago Press.

Carroll, J. B. 1995, "Reflections on Stephen Jay Gould's *The Mismeasure of Man*," *Intelligence* 21: 121–134.

Caryl, P. G. and Deary, I. J. 1996, "IQ and Censorship," *Nature* 381: 270.

Caspi, A., McClay, J., Moffitt, T. E., and Mill, J. 2002, "Role of Genotype in the Cycle of Violence in Maltreated Children," *Science* 297: 851–854.

Cattell, R. B. 1963, "The Interaction of Hereditary and Environmental Influences," *British Journal of Statistical Psychology* 16: 191–210.

1965, "Methodological and Conceptual Advances in Evaluating Hereditary and Environmental Influences and their Interaction," in S. G. Vandenberg (ed.), *Methods and Goals in Human Behavior Genetics*, New York, Academic Press.

1982, *The Inheritance of Personality and Ability*, New York, Academic Press.

Cohen, J. 1988, *Statistical Power Analysis for the Behavioral Sciences*, 2nd edn., Hillsdale, N.J., Lawrence Erlbaum.

Collins, F. S., Weiss, L., and Hudson, K. 2001, "Have No Fear, Genes Aren't Everything: Heredity and Humanity," *New Republic Online*, June 25.

Cooper, R. M. and Zubek, J. P. 1958, "Effects of Enriched and Restricted Early Environments on the Learning Ability of Bright and Dull Rats," *Canadian Journal of Psychology* 12: 159–164.

Cosmides, L. and Tooby, J. 1997, "Evolutionary Psychology: A Primer," http://www.psych.ucsb.edu/research/cep/primer.html, accessed June 2003.

Crawford, C. B. 2003, "A Prolegomenon for a Viable Evolutionary Psychology: The Myth and the Reality," *Psychological Bulletin* 129: 854–857.

Crow, J. F. 1969, "Genetic Theories and Influences: Comments on the Values of Diversity," in *Environment, Heredity and Intelligence*, Reprint Series No. 2 compiled from the *Harvard Educational Review*, Cambridge, Mass., Harvard Education Publishing Group.

Cummings, P. W. 1967, "Racism," in P. Edwards (ed.), *Encyclopedia of Philosophy*, vol. 7, New York, Free Press.

Daniels, M., Devlin, B., and Roeder, K. 1997, "Of Genes and IQ," in B. Devlin, S. E. Fienberg, D. P. Resnick, and K. Roeder (eds.), *Intelligence, Genes and Success: Scientists Respond to the Bell Curve*, New York, Springer.

Daniels, N. 1976, "IQ, Heritability and Human Nature," in R. S. Cohen and A. C. Michalos (eds.), *PSA 1974: Proceedings of the Philosophy of Science Association*, Dordrecht, Reidel.

Davis, B. D. 1986, *Storm over Biology: Essays on Science, Sentiment, and Public Policy*, Buffalo, N.Y., Prometheus Books.

Dawkins, R. 1982, *The Extended Phenotype: The Long Reach of the Gene*, Oxford, Oxford University Press.

1985, "Sociobiology: The Debate Continues," *New Scientist*, January 24.

1998, "Genes and Determinism" (interview), http://www.world-of-dawkins.com/Dawkins/Work/Interviews/genes_and_determinism.shtml, accessed September 1, 2004.

2003, *A Devil's Chaplain*, London, Weidenfeld & Nicolson.

De Jong, H. L. 2000, "Genetic Determinism: How Not to Interpret Behavior Genetics," *Theory & Psychology* 10: 615–637.

De Waal, F. B. M. 1999, "The End of Nature versus Nurture," *Scientific American* 281: 94–99.

Deary, I. J. 1998, "Differences in Mental Abilities," *British Medical Journal* 317: 1701–1703.

2000, *Looking Down on Human Intelligence: From Psychometrics to the Brain*, Oxford, Oxford University Press.

2001, "Individual Differences in Cognition: British Contributions Over a Century," *British Journal of Psychology* 92: 217–237.

DeFries, J. C. 1967, "Quantitative Genetics and Behavior: Overview and Perspective," in J. Hirsch (ed.) *Behavior-Genetic Analysis*, New York, McGraw-Hill.

1972, "Quantitative Aspects of Genetics and Environment in the Determination of Behavior," in L. Ehrman, G. S. Omenn, and E. Caspari (eds.), *Genetics, Environment and Behavior*, New York, Academic Press.

Dennett, D. C. 1995, *Darwin's Dangerous Idea*, New York, Simon & Schuster.

2003a, "The Mythical Threat of Genetic Determinism," *Chronicle of Higher Education*, January 31.

2003b, *Freedom Evolves*, New York, Viking.

Detterman, D. K. 1990, "Don't Kill the ANOVA Messenger for Bearing Bad Interaction News," *Behavioral and Brain Sciences* 13: 131–132.

2000, "General Intelligence and the Definition of Phenotypes," in G. Bock, J. Goode, and K. Webb (eds.), *The Nature of Intelligence*, New York, Wiley.

Diamond, J. 1997, *Guns, Germs and Steel: The Fates of Human Societies*, New York, Norton.

Dickens, W. T. and Flynn, J. R. 2001, "Heritability Estimates versus Large Environmental Effects: The IQ Paradox Resolved," *Psychological Review* 108: 346–369.

Dobzhansky, T. 1955, *Evolution, Genetics, and Man*, New York, John Wiley.

1956, *The Biological Basis of Human Freedom*, New York, Columbia University Press.

1973, *Genetic Diversity and Human Equality*, New York, Basic Books.

Downes, S. M. 2004, "Heredity and Heritability," *Stanford Encyclopedia of Philosophy*, http://plato.stanford.edu/entries/heredity/, accessed July 19, 2004.

Dummett, M. 1981, "Ought Research to Be Unrestricted?" *Grazer Philosophische Studien* 12–13: 281–298.

Dupré, J. 1995, *The Disorder of Things*, Cambridge, Mass., Harvard University Press.

2001, *Human Nature and the Limits of Science*, Oxford, Clarendon Press.

2003, *Darwin's Legacy: What Evolution Means Today*, Oxford, Oxford University Press.

2004, "Science and Values and Values in Science: Comments on Philip Kitcher's *Science, Truth and Democracy*," *Inquiry* 47: 505–514.

Eaves, L. J., Last, K., Martin, N. G., and Jinks, J. L. 1977, "A Progressive Approach to Non-Additivity and Genotype-Environmental Covariance in the Analysis of Human Differences," *British Journal of Mathematical and Statistical Psychology* 30: 1–42.

Edwards, A. W. F. 2003, "Human Genetic Diversity: Lewontin's Fallacy," *BioEssays* 25: 798–801.

Ehrlich, P. 2000, "The Tangled Skeins of Nature and Nurture in Human Evolution," *Chronicle of Higher Education*, September 22.

Ehrlich, P. and Feldman, M. W. 2003, "Genes and Cultures: What Creates our Behavioral Phenome?" *Current Anthropology* 44: 87–95.

Eisenberg, L. 1995, "The Social Construction of the Human Brain," *American Journal of Psychiatry* 152: 1563–1575.

2001, "Why Has the Relationship between Psychiatry and Genetics Been So Contentious?" *Genetics in Medicine* 3: 377–381.

2002, "Is Biology Destiny? Is It All in Our Genes?" *Journal of Psychiatric Practice* 8: 337–343.

Emigh, T. H. 1977, "Partition of Phenotypic Variance under Unknown Dependent Association of Genotypes and Environments," *Biometrics* 33: 505–514.

Etzioni, A. 1973, *Genetic Fix: The Next Technological Revolution*, New York, Harper & Row.

Eysenck, H. J. 1985, *Personality and Individual Differences: A Natural Science Approach*, New York, Plenum Press.

1994, "The Impact of Genetic Factors on Behavior," *Personality and Individual Differences* 17: 581–582.

1998, *Intelligence: A New Look*, New Brunswick, N.J., Transaction.

Falconer, D. S. 1989, *Introduction to Quantitative Genetics*, 3rd edn., Harlow, Essex, Longman.

Falk, R. 2001, "Can the Norm of Reaction Save the Gene Concept?" in R. S. Singh, K. Krimbas, D. B. Paul, and J. Beatty (eds.), *Thinking About Evolution: Historical, Philosophical and Political Perspectives*, Cambridge, Cambridge University Press.

Fausto Sterling, A. 1992, *Myths of Gender*, New York, Basic Books.

Feingold, A. 1992, "Good-looking People Are Not What We Think," *Psychological Bulletin* 111: 304–341.

Feldman, M. W. 1992, "Heritability: Some Theoretical Ambiguities," in E. F. Keller and E. A. Lloyd (eds.), *Keywords in Evolutionary Biology*, Cambridge, Mass., Harvard University Press.

2001, "The Meaning of Race: Genes, Environments and Affirmative Action," *Berkeley La Raza Law Journal* 12: 365–371.

Feldman, M. W. and Lewontin, R. C. 1975, "The Heritability Hang-up," *Science* 190: 1163–1168.

"Response to Letters on 'Heritability Hang-Up,'" *Science* 194: 12–14.

Fine, A. 1990, "Causes of Variability: Disentangling Nature and Nurture," *Midwest Studies in Philosophy* 15: 94–113.

1991, "Piecemeal Realism," *Philosophical Studies* 61: 79–96.

Fisher, R. A. 1930, *The Genetical Theory of Natural Selection*, Oxford, Clarendon Press.

1951, "Limits to Intensive Production in Animals," *British Agricultural Bulletin* 4: 217–218.

1959, *Statistical Methods and Scientific Inference*, 2nd edn., Edinburgh, Oliver and Boyd.

Fletcher, R. 1991, *Science, Ideology, and the Media: The Cyril Burt Scandal*, New Brunswick, N.J., Transaction.

Flynn, J. R. 1980, *Race, IQ and Jensen*, London, Routledge & Kegan Paul.

1987, "Race and IQ: Jensen's Case Refuted," in S. Modgil and C. Modgil (eds.), *Arthur Jensen: Consensus and Controversy*, London, Falmer Press.

1989, "Rushton, Evolution, and Race: An Essay on Intelligence and Race," *Psychologist* 2: 363–366.

1999a, "Evidence Against Rushton: The Genetic Loading of WISC-R Subtests and the Causes of Between-Group Differences," *Personality and Individual Differences* 26: 373–379.

1999b, "Searching for Justice: The Discovery of IQ Gains over Time," *American Psychologist* 54: 5–19.

Fulker, D. W. 1974, "Applications of Biometrical Genetics to Human Behavior," in J. H. F. van Abeleen (ed.), *The Genetics of Behavior*, Amsterdam, North-Holland.

1975, Review of L. J. Kamin's *The Science and Politics of IQ*, *American Journal of Psychology* 88: 505–519.

Fuller, J. L. 1972, "Discussion," in L. Ehrman, G. S. Omenn, and E. Caspari (eds.), *Genetics, Environment and Behavior*, New York, Academic Press.

1979, "Comment on Wahlsten," in J. R. Royce and L. P. Mos (eds.), *Theoretical Advances in Behavior Genetics*, Alphen aan den Rijn, Sijthoff & Noordhof.

Fuller, J. L. and Thompson, W. R. 1960, *Behavior Genetics*, New York, John Wiley.

Galton, F. 1874, *English Men of Science: Their Nature and Nurture*, London, Macmillan.

1889, *Natural Inheritance*, London, Macmillan.

1892, *Hereditary Genius*, 2nd edn., London, Macmillan.

Garfield, E. 1978, "High Impact Science and the Case of Arthur Jensen," *Current Contents* 41: 652–662.

Garfinkel, A. 1981, *Forms of Explanation: Rethinking the Questions in Social Theory*, New Haven, Conn., Yale University Press.

Genovese, E. 1995, "Living with Inequality," in R. Jacoby and N. Glauberman (eds.), *The Bell Curve Debate: History, Documents, Opinions*, New York, Random House.

Gibbard, A. 2001, "Genetic Plans, Genetic Differences, and Violence: Some Chief Possibilities," in D. Wasserman and R. Wachbroit (eds.), *Genetics and Criminal Behavior*, Cambridge, Cambridge University Press.

Gifford, F. 1990, "Genetic Traits," *Biology and Philosophy* 5: 327–347.

Gigerenzer, G. 1997, "Tools," in P. Weingart, S. Maasen, P. J. Richerson, and S. D. Mitchell (eds.), *Human by Nature: Between Biology and Social Sciences*, Mahwah, N.J., Erlbaum.

Gilbert, S. 2002, "Genetic Determinism: The Battle Between Scientific Data and Social Image in Contemporary Developmental Biology," in A. Grunwald, M. Gutman, and E. M. Neumann-Held (eds.), *On Human Nature: Anthropological, Biological and Philosophical Foundations*, Berlin, Springer.

Glazer, N. 1994, "Lying Game," *New Republic*, October.

Glover, J. 2001, "Future People, Disability, and Screening," in J. Harris (ed.), *Bioethics*, Oxford, Oxford University Press.

Glymour, C. 1997, "Social Statistics and Genuine Inquiry: Reflections on *The Bell Curve*," in B. Devlin, S. E. Fienberg, D. P. Resnick, and K. Roeder (eds.), *Intelligence, Genes, and Success: Scientists Respond to The Bell Curve*, New York, Springer.

1998, "What Went Wrong? Reflections on Science by Observation and *The Bell Curve*," *Philosophy of Science* 65: 1–32.

Godfrey-Smith, P. 1998, *Complexity and the Function of Mind in Nature*, Cambridge, Cambridge University Press.

1999, "Genes and Codes: Lessons from the Philosophy of Mind?" in V. Hardcastle (ed.), *Where Biology Meets Psychology*, Cambridge, Mass., MIT Press.

2000, "On the Theoretical Role of 'Genetic Coding'," *Philosophy of Science* 67: 26–44.

2002, "Goodman's Problem and Scientific Methodology," *Journal of Philosophy* 100: 573–590.

Goldsmith, H. H. 1993, "Nature–Nurture Issues in the Behavioral Genetics Context: Overcoming Barriers to Communication," in R. Plomin and G. E. McClearn (eds.), *Nature, Nurture and Psychology*, Washington, D.C., American Psychological Association.

Gottesman, I. I. 2002, Foreword, in J. Benjamin, R. P. Ebstein, and R. H. Belmaker (eds.), *Molecular Genetics and Human Personality*, Washington, D.C., American Psychiatric Publishing, Inc.

Gottfredson, L. S. 1994, "Egalitarian Fiction and Collective Fraud," *Society* 31: 53–59.

2002, "Where and Why g Matters: Not a Mystery," *Human Performance* 15: 25–46.

2005, "Suppressing Intelligence Research: Hurting Those We Intend to Help," in R. H. Wright, and N. A. Cummings (eds.), *Destructive Trends in Mental Health: The Well Intentioned Path to Harm*, London, Brunner-Routledge.

Gottfredson, L. S., Arvey R. D., Bouchard, T. J., Jr., et al. 1994, "Mainstream Science on Intelligence," *Wall Street Journal*, December 13.

Gottlieb, G. 1992, *Individual Development and Evolution: The Genesis of Novel Behavior*, New York, Oxford University Press.

1995, "Some Conceptual Deficiencies in 'Developmental' Behavior Genetics," *Human Development* 38: 131–141.

2001, "Nature and Nurture Theories," in A. E. Kazdin (ed.), *Encyclopedia of Psychology*, Oxford, Oxford University Press.

2003a, "On Making Behavioral Genetics Truly Developmental," *Human Development* 46: 337–355.

2003b, "Probabilistic Epigenesis of Development," in J. Valsiner and K. J. Connoly (eds.), *Handbook of Developmental Psychology*, London, Sage.

2004, "Normally Occurring Environmental and Behavioral Influences on Gene Activity," in C. G. Coll, E. L. Bearer, and R. M. Lerner. (eds.), *Nature and Nurture*, Mahwah, N.J., Lawrence Erlbaum.

Gould, S. J. 1977, *Ever Since Darwin: Reflections in Natural History*, Harmondsworth, Penguin.

1981, *The Mismeasure of Man*, Harmondsworth, Penguin.

1983, *The Panda's Thumb*, Harmondsworth, Penguin.

1995, "Ghosts of Bell Curves Past," *Natural History* 104: 12–17.

Graves, R. L. 2001, *The Emperor's New Clothes: Biological Theories of Race at the Millennium*, New Brunswick, N.J., Rutgers University Press.

References

Gray, J. R. and Thompson, P. M. 2004, "Neurobiology of Intelligence: Science and Ethics," *Nature Reviews Neuroscience* 5: 471–482.

Gray, R. D. 1992, "Death of the Gene: Developmental Systems Strike Back," in P. E. Griffiths (ed.), *Trees of Life: Essays in Philosophy of Biology*, Dordrecht, Kluwer.

2001, "Selfish Genes or Developmental Systems," in S. Oyama, P. E. Griffiths, and R. D. Gray (eds.), *Cycles of Contingency: Developmental Systems and Evolution*, Cambridge, Mass., MIT Press.

Griffiths, A. J. F., Miller, J. H., Suzuki, D. T., and Lewontin, R. C. 2000, *Introduction to Genetic Analysis*, 7th edn., New York, Freeman.

Griffiths, P. E. 2002a, "The Fearless Vampire Conservator: Philip Kitcher, Genetic Determinism and the Informational Gene," http://philsci-archive. pitt.edu/archive/00000728/00/fearless_vampire_conservator.pdf, accessed March 2004.

2002b, "The Fearless Vampire Conservator: Philip Kitcher and Genetic Determinism," http://philsci-archive.pitt.edu/ archive/00000652/00/vampire_slayer_final.pdf, accessed March 2004.

Griffiths, P. E. and Gray, R. D. 1994, "Developmental Systems and Evolutionary Explanation," *Journal of Philosophy* 91: 277–304.

Griffiths, P. E. and Knight, R. D. 1998, "What Is the Developmentalist Challenge?" *Philosophy of Science* 65: 253–258.

Grove, V. 1991, "To Life's Frontier with the Doctor in Genes," *Sunday Times* (London), November 17.

Guo, S.-W. 1999, "The Behaviors of Some Heritability Estimators in the Complete Absence of Genetic Factors," *Human Heredity* 49: 215–228.

2000, "Gene–Environment Interaction and the Mapping of Complex Traits," *Human Heredity* 50: 286–303.

Gutmann, A. 1980, *Liberal Equality*, Cambridge, Cambridge University Press.

Haack, S. 1996, "Concern for Truth: What it Means, Why it Matters," in P. R. Gross, N. Levitt, and M. W. Lewis (eds.), *The Flight from Science and Reason*, New York, New York Academy of Sciences.

2003, *Defending Science – Within Reason*, New York, Prometheus Books.

Hacking, I. 1995, "Pull the Other One," *London Review of Books*, January 26.

Haldane, J. B. S. 1938, *Heredity and Politics*, London, George Allen & Unwin.

Halpern, D. F., Gilbert, R., and Coren, S. 1996, "PC or Not PC? Contemporary Challenges to Unpopular Research Findings," *Journal of Social Distress and the Homeless* 5: 251–271.

Harpending, H. C. 1995, "Human Biological Diversity," *Evolutionary Anthropology* 4: 99–102.

2002, "Race: Population Genetics Perspectives," in M. Pagel (ed.), *Encyclopedia of Evolution*, Oxford, Oxford University Press.

Harris, J. R. 1996, "A Defense of Behavioral Genetics by a Non-Behavioral Geneticist," *Psycholoquy*, May 12.

1998, *The Nurture Assumption: Why Children Turn Out the Way They Do*, New York, Free Press.

Havender, W. R. 1976, "Discussion," *Science* 194: 608–609.

Hay, D. A. 1985, *Essentials of Behavior Genetics*, Melbourne, Blackwell.

Hayek, F. A. 1960, *The Constitution of Liberty*, Chicago, University of Chicago Press.

1978, "Nature v. Nurture Once Again," in   F. A. Hayek (ed.), *New Studies in Philosophy, Politics, Economics and the History of Ideas*, London, Routledge & Kegan Paul.

Hebb, D. O. 1970, "A Return to Jensen and his Social Critics," *American Psychologist* 25: 568–587.

1980, *Essay on Mind*, Hillsdale, N.J., Lawrence Erlbaum.

Hellevik, O. 1984, *Introduction to Causal Analysis: Exploring Survey Data by Crosstabulation*, London, George Allen & Unwin.

Herbert, W. 1997, "Politics of Biology," *U.S. News & World Report*, April 21, 1997.

Herrnstein, R. J. 1971, "IQ," *Atlantic Monthly*, September.

1973, *I.Q. in the Meritocracy*, Boston, Mass., Atlantic Monthly Press.

Herrnstein, R. J. and Murray, C. 1994, *The Bell Curve: Intelligence and Class Structure in American Life*, New York, Free Press.

Hettema, J. M., Neale, M. C., and Kendler, K. S. 1995, "Physical Similarity and the Equal Environment Assumption in Twin Studies of Psychiatric Disorders," *Behavior Genetics* 25: 327–335.

Hirsch, J. 1976, "Behavior-Genetic Analysis and Its Biosocial Consequences," in N. J. Block and G. Dworkin (eds.), *The IQ Controversy*, New York, Pantheon.

1990, "A Nemesis for Heritability Estimation," *Behavioral and Brain Sciences* 13: 137–138.

1997, "Some History of Heredity-vs-Environment, Genetic Inferiority at Harvard(?), and the (Incredible) Bell Curve," *Genetica* 99: 207–224.

Hitchcock, C. 2003, Review of Clark Glymour's *The Mind's Arrows*, *Mind* 112: 340–343.

Hocutt, M. and Levin, M. 1999, "The *Bell Curve* Case for Heredity," *Philosophy of the Social Sciences* 29: 389–415.

Hogben, L. 1933, *Nature and Nurture*, London, George Allen & Unwin.

Hubbard, R. and Wald, E. 1993, *Exploding the Gene Myth*, Boston, Mass., Beacon Press.

Hume, D. 1999, *An Enquiry Concerning Human Understanding*, Oxford, Oxford University Press.

Jacquard, A. 1983, "Heritability: One Word, Three Concepts," *Biometrics* 39: 465–477.

1984, *In Praise of Difference: Genetics and Human Affairs*, New York, Columbia University Press.

1985, *Endangered by Science*, New York, Columbia University Press.

Jencks, C. 1980, "Heredity, Environment, and Human Policy Reconsidered," *American Sociological Review* 45: 723–736.

1988, "Whom Must We Treat Equally for Educational Opportunity to Be Equal?," *Ethics* 98: 518–533.

1992, *Rethinking Social Policy*, Cambridge, Mass., Harvard University Press.

Jencks, C. and Bane, M. J. 1976, "Five Myths about your IQ," in N. J. Block and G. Dworkin (eds.), *The IQ Controversy: Critical Readings*, New York, Pantheon.

Jencks, C. S., Smith, M., Acland, H., et al. 1972, *Inequality. A Reassessment of the Effect of Family and Schooling in America*, New York, Basic Books.

Jensen, A. R. 1968, "Another Look at Culture-Fair Tests," Western Regional Conference on Testing Problems, Berkeley, Calif., Educational Testing Service, Western Office.

1969a, Reducing the Heredity–Environment Uncertainty, in *Environment, Heredity and Intelligence*, Reprint Series No. 2 compiled from the *Harvard Educational Review*, Cambridge, Mass., Harvard Education Publishing Group.

1969b, "How Much Can We Boost I.Q. and Scholastic Achievement?" *Harvard Educational Review* 39: 1–123.

1971, "Hebb's Confusion about Heritability," *American Psychologist* 26: 394.

1972a, *Genetics and Education*, London, Methuen.

1972b, "Discussion," in L. Ehrman, G. S. Omenn, E. Caspari (eds.), *Genetics, Environment and Behavior*, New York, Academic Press.

1972c, "Interpretation of Heritability," *American Psychologist* 27: 973–975.

1973a, *Educability and Group Differences*, New York, Harper & Row.

1973b, *Educational Differences*, London, Methuen.

1976a, "The Problem of Genotype–Environment Correlation in the Estimation of Heritability from Monozygotic and Dizygotic Twins," *Acta Geneticae Medicae et Gemellologiae* 25: 86–99.

1976b, "Race and Genetics of Intelligence: A Reply to Lewontin," in N. J. Block and G. Dworkin (eds.), *The IQ Controversy*, New York, Pantheon.

1977, "Race and Mental Ability," in A. H. Halsey (ed.), *Heredity and Environment*, London, Methuen.

1980, *Bias in Mental Testing*, New York, Free Press.

1981a, *Straight Talk About Mental Tests*, New York, Free Press.

1981b, "Obstacles, Problems, and Pitfalls in Differential Psychology," in S. W. Scarr (ed.), *Race, Social Class and Individual Differences in I.Q.: New Studies of Old Issues*, Hillsdale, N.J., Erlbaum.

1982, "The Debunking of Scientific Fossils and Straw Persons," *Contemporary Education Review* 1: 121–135.

1991, "Spearman's g and the Problem of Educational Equality," *Oxford Reviews of Education* 17: 169–185.

1994, "Race and IQ Scores," in R. J. Sternberg (ed.), *Encyclopedia of Human Intelligence*, New York, Macmillan.

1998, *The g Factor: The Science of Mental Ability*, Westport, Conn., Praeger.

Jinks, J. L. and Fulker, D. W. 1970, "Comparison of the Biometrical Genetical, MAVA, and Classical Approaches to the Analysis of Human Behavior," *Psychological Bulletin* 73: 311–349.

Johansson, I. 1961, *Genetic Aspects of Dairy Cattle Breeding*, Edinburgh, Oliver and Boyd.

Johnston, T. D. 1987, "The Persistence of Dichotomies in the Study of Behavioral Development," *Developmental Review* 7: 149–182.

2003, "Comments on Ehrlich & Feldman 2003," *Current Anthropology* 44: 99.

Johnston, T. D. and Edwards, L. 2002, "Genes, Interactions, and the Development of Behavior," *Psychological Review* 109: 26–34.

Jones, S. 1994a, *The Language of the Genes*, London, Flamingo.

1994b, "Our Favorite Genes," *Newsweek*, December 26–January 2.

1996, "Don't Blame the Genes," *Guardian* (London), June 7.

Jones, S. and Van Loon, B. 1993, *Genetics for Beginners*, Cambridge, Icon Books.

Joynson, R. B. 1989, *The Burt Affair*, London, Routledge.

2003, "Selective Interest and Psychological Practice: A New Interpretation of the Burt Affair," *British Journal of Psychiatry* 94: 409–426.

Kagan, J. 1998, *Three Seductive Ideas*, Cambridge, Mass., Harvard University Press.

Kagan, J. and Snidman, N. 2004, *The Long Shadow of Temperament*, Cambridge, Mass., Harvard University Press.

Kamin, L. J. 1974, *The Science and Politics of I.Q.*, Potomac, Md., Lawrence Erlbaum.

Kanazawa, S. and Kovar, J. L. 2004, "Why Beautiful People are More Intelligent," *Intelligence* 32: 227–243.

Kaplan, J. M. 2000, *The Limits and Lies of Human Genetic Research*, New York & London, Routledge.

Kassim, H. 2002, "'Race,' Genetics, and Human Difference," in J. Burley and J. Harris (eds.), *A Companion to Genetics*, Oxford, Blackwell.

Keller, E. F. 1994, "Rethinking the Meaning of Genetic Determinism," in *The Tanner Lectures on Human Values*, 15, Salt Lake City, University of Utah Press.

2001, "Beyond the Gene but Beneath the Skin," in S. Oyama, P. E. Griffiths, and R. D. Gray (eds.), *The Cycles of Contingency: Developmental Systems and Evolution*, Cambridge, Mass., MIT Press.

Kempthorne, O. 1997, "Heritability: Uses and Abuses," *Genetica* 99: 109–112.

Kendler, K. S., Neale, M. C., Kessler, R. C., Heath, A. C., and Eaves, L. J. 1993, "A Test of the Equal Environment Assumption in Twin Studies of Psychiatric Illness," *Behavior Genetics* 23: 21–27.

Kendler, K. S. and Prescott, C. A. 1999, "Caffeine Intake, Tolerance, and Withdrawal in Women: A Population-Based Study," *American Journal of Psychiatry* 156: 223–228.

Kendler, K. S., Thornton, L. M., Gilman, S. E., and Kessler, R. O. 2000, "Sexual Orientation in a U.S. National Sample of Twin and Nontwin Sibling Pairs," *American Journal of Psychiatry* 157: 1843–1846.

Khoury, M. J. and Thornburg, R. S. 2001, "Nature versus Nurture: An Unnecessary Debate," *Gene Letter*, March 1.

Kitcher, P. 1984, "The Case Against Sociobiology," *New York Times Book Review*, May 20.

1985, *Vaulting Ambition: Sociobiology and the Quest for Human Nature*, Cambridge, Mass., MIT Press.

1987, "The Transformation of Human Sociobiology," in A. Fine and P. Machamer (eds.), *PSA 1986: Proceedings of the Philosophy of Science Association*, East Lansing, Mich., Philosophy of Science Association.

1989, Review of R. Levins and R. Lewontin, *The Dialectical Biologist*, *Philosophical Review* 98: 262–266.

1990, "Developmental Decomposition and the Future of Human Behavioral Ecology," *Philosophy of Science* 57: 96–117.

1996, *The Lives to Come: The Genetic Revolution and Human Possibilities*, New York, Simon & Schuster.

1997, "An Argument about Free Inquiry," *Nous* 31: 279–306.

1998, "A Plea for Science Studies," in, N. Koertge (ed.), *A House Built on Sand: Exposing Postmodernist Myths About Science*, New York, Oxford University Press.

1999, "Race, Ethnicity, Biology, Culture," in L. Harris (ed.), *Racism*, Amherst, Mass., Humanity Books.

2001a, *Science, Truth, and Democracy*, Oxford, Oxford University Press.

2001b, "Battling the Undead: How (and How Not) to Resist Genetic Determinism," in R. S. Singh, K. Krimbas, D. B. Paul, and J. Beatty (eds.), *Thinking about Evolution: Historical, Philosophical and Political Perspectives*, Cambridge, Cambridge University Press.

2004, "Evolutionary Theory and the Social Uses of Biology," *Biology and Philosophy* 19: 1–15.

Klump, K. L., Holly, A., Iacono, W. G., McGue, M., and Willson, L. E. 2000, "Physical Similarity and Twin Resemblance for Eating Attitudes and Behaviors: A Test of the Equal Environments Assumption," *Behavior Genetics* 30: 51–58.

Koehler, J. J. 1996, "The Base Rate Fallacy Reconsidered: Descriptive, Normative, and Methodological Challenges," *Behavioral and Brain Sciences* 19: 1–53.

Kurzban, R. 2004, "Nibbling on Nature and Nurture," *Trends in Ecology & Evolution* 19: 290–291.

Ladd, E. C. and Lipset, S. M. 1975, *The Divided Academy: Professors and Politics*, New York, McGraw-Hill.

Langlois, J. H., Kalakanis, L., Rubenstein, A. J., Larson, A., Hallam, M., and Smoot, M. 2000, "Maxims or Myths of Beauty? A Meta-Analytic and Theoretical Review," *Psychological Bulletin* 126: 390–423.

Layzer, D. 1972a, "Science or Superstition? (A Physical Scientist Looks at the IQ Controversy)," *Cognition* 1: 265–299.

1972b, "A Rejoinder to Professor Herrnstein's Comments," *Cognition* 1: 423–426.

1974, "Heritability Analyses of IQ Scores: Science or Numerology," *Science* 183: 1259–1266.

1976, "Science or Superstition? A Physical Scientist Looks at the IQ Controversy" (a revised version of Layzer 1972a), in N. J. Block and G. Dworkin (eds.), *The IQ Controversy: Critical Readings*, New York, Pantheon.

LeDoux, J. E. 1998, "Nature vs. Nurture: The Pendulum Still Swings with Plenty of Momentum," *Chronicle of Higher Education*, December 11.

Lehrman, D. S. 1970, "Semantic and Conceptual Issues in the Nature–Nurture Problem," in L. R. Aronson, D. S. Lehrman, E. Tobach, and J. S. Rosenblatt (eds.), *Development and Evolution of Behavior*, San Francisco, Calif., W. H. Freeman.

Lerner, R. M. 1972, "Polygenic Inheritance and Human Intelligence," *Evolutionary Biology* 6: 399–414.

1986, *Concepts and Theories of Human Development*, 2nd edn., New York, Random House.

1992, *Final Solutions: Biology, Prejudice, and Genocide*, University Park, Pennsylvania State University Press.

Levin, M. 1994, "A Formal Treatment of Gene Identity, Genetic Causation, and Related Notions," *Behavior and Philosophy* 22: 49–58.

1997, *Why Race Matters*, New York, Praeger.

Lewontin, R. C. 1972, "The Apportionment of Human Diversity," *Evolutionary Biology* 6: 381–398.

1973, "Herrnstein's Sleight-of-Hand," *Harvard Crimson*, December 13.

1975, "Genetic Aspects of Intelligence," *Annual Review of Genetics* 9: 387–405.

1976a, "The Analysis of Variance and the Analysis of Causes," in N. J. Block and G. Dworkin (eds.), *The IQ Controversy*, New York, Pantheon.

1976b, "The Fallacy of Biological Determinism," *The Sciences* 16: 6–10.

1976c, "Race and Intelligence," in N. J. Block and G. Dworkin (eds.), *The IQ Controversy*, New York, Pantheon.

1976d, Review of L. J. Kamin's *The Science and Politics of IQ*, *Contemporary Psychology* 21: 97–98.

1977, "Biological Determinism as a Social Weapon," in SESPA (ed.), *Biology as a Social Weapon*, Minneapolis, Burgess.

1982, *Human Diversity*, New York, Scientific American Books.

1983, "Gene, Organism and Environment," in D. S. Bendall (ed.), *Evolution from Molecules to Men*, Cambridge, Cambridge University Press.

1993, *Biology As Ideology: The Doctrine of DNA*, New York, Harper/Collins.

2000, *The Triple Helix: Gene, Organism, and Environment*, Cambridge, Mass., Harvard University Press.

2003, PBS interview (edited transcript), http://www.pbs.org/race /000_about/ 002_04-background-01–04.htm, accessed June 25, 2004.

2004, "The Genotype/Phenotype Distinction," *Stanford Encyclopedia of Philosophy*, http://plato.stanford.edu/archives/ spr2004/entries/genotype-phenotype, accessed August 2004.

Lilienfeld, S. and Waldman, I. 2000, "Race and IQ: What Science Says," *Emory Report*, 13 November.

Locke, J. 1959, *An Essay Concerning Human Understanding*, New York, Dover.

Loehlin, J. C. 1992, *Genes and Environment in Personality Development*, London, Sage.

2001, "Behavior Genetics and Parenting Theory," *American Psychologist* 56: 169–170.

2002, "The IQ Paradox Resolved? Still an Open Question," *Psychological Review* 109: 754–758.

Loehlin, J. C. and DeFries, J. D. 1987, "Genotype–Environment Correlation and IQ," *Behavior Genetics* 17: 263–277.

Loehlin, J. C., Lindzey, G., and Spuhler, J. N. 1975, *Race Differences in Intelligence*, San Francisco, Calif., W. H. Freeman.

Loehlin, J. C. and Nichols, R. C. 1976, *Heredity, Environment and Personality: A Study of 850 Sets of Twins*, Austin, University of Texas Press.

Longino, H. E. 1990, *Science as Social Knowledge*, Princeton, Princeton University Press.

Lubinski, D. 2004, "Introduction to the Special Section on Cognitive Abilities," *Journal of Personality and Social Psychology* 86: 96–111.

Luria, S. E. 1999, "What Can Biologists Solve?" in A. Montagu (ed.), *Race and IQ*, New York, Oxford University Press.

Lykken, D. 1995, *The Antisocial Personalities*, Hillsdale, N.J., Lawrence Erlbaum.

 1998a, "The Genetics of Genius," in A. Steptoe (ed.), *Genius and the Mind: Studies of Creativity and Temperament in the Historical Record*, Oxford, Oxford University Press.

 1998b, "How Can Educated People Continue to Be Radical Environmentalists?" http://www.edge.org/documents/archive/edge43.html, accessed July 17, 2003.

Lynn, R. 1982, "IQ in Japan and the United States Show a Growing Disparity," *Nature* 297: 222–223.

Mackintosh, N. J. 1995a, "Declining Educational Standards," in N. J. Mackintosh (ed.), *Cyril Burt: Fraud or Framed*, Oxford, Oxford University Press.

 1995b, "Twins and Other Kinship Studies," in N. J. Mackintosh (ed.), *Cyril Burt: Fraud or Framed*, Oxford, Oxford University Press.

 1998, *IQ and Human Intelligence*, Oxford, Oxford University Press.

Maclaurin, J. 2002, "The Resurrection of Innateness," *Monist* 85: 105–130.

Mazur, A. and Robertson, L. S. 1972, *Biology and Social Behavior*, New York, Free Press.

McCall, R. M. 1991, "So Many Interactions, So Little Evidence. Why?" in T. D. Wachs and R. Plomin (eds.), *Conceptualization and Measurement of Organism–Environment Interaction*, Washington, D.C., American Psychological Association.

McClearn, G. E. 1963, "Genetics of Behavior," in L. Postman (ed.), *Psychology in the Making*, New York, Alfred Knopf.

 1964, "The Inheritance of Behavior," in L. Postman (ed.), *Psychology in the Making*, New York, Alfred Knopf.

McClelland, G. H. and Judd, C. M. 1993, "Statistical Difficulties of Detecting Interactions and Moderator Effects," *Psychological Bulletin* 114: 376–390.

McGue, M. 1989, "Nature–Nurture and Intelligence," *Nature* 340: 507–508.

McGue, M. and Bouchard, T. J., Jr. 1998, "Genetic and Environmental Influences on Human Behavioral Differences," *Annual Reviews of Neuroscience* 21: 1–24.

McGuffin, P. 2000, "The Quantitative and Molecular Genetics of Human Intelligence," in G. Bock, J. Goode, and K. Webb (eds.), *The Nature of Intelligence*, New York, Wiley.

McGuffin, P. and Katz, R. 1990, "Who Believes in Estimating Heritability as an End in Itself?" *Behavioral and Brain Sciences* 13: 141–142.

McGuffin, P. and Martin, N. 1999, "Behaviour and Genes," *British Medical Journal* 319: 37–40.

McInerney, J. D. 1996, "Why Biological Literacy Matters: A Review of Commentaries Related to *The Bell Curve*," *Quarterly Review of Biology* 71: 81–96.

Meaney, M. J. 2001, "Nature, Nurture, and the Disunity of Knowledge," *Annals of the New York Academy of Sciences* 935: 50–61.

Medawar, P. 1977a, "Unnatural Science," *New York Review of Books*, February 3.

1977b, "Unnatural Science Cont'd," *New York Review of Books*, June 23.

Meehl, P. E. 1970, "Nuisance Variables and the Ex Post Facto Design," in M. Radner and S. Winokur (eds.), *Analyses of Theories and Methods of Physics and Psychology*, Minnesota Studies in the Philosophy of Science, vol. 4, Minneapolis, University of Minnesota Press.

1998, "The Power of Quantitative Thinking," http://www.tc.umn.edu/~pemeehl/ PowerQuantThinking.pdf, accessed August 2004.

Michael, J. S. 1988, "A New Look at Morton's Craniological Research," *Current Anthropology* 29: 349–354.

Michel, G. F. and Moore, C. L. 1995, *Developmental Psychobiology: An Interdisciplinary Science*, Cambridge, Mass., MIT Press.

Miele, F. 2002, *Intelligence, Race, and Genetics*, Boulder, Colo., Westview Press.

Millstein, R. L. 2002, "Evolution," in P. Machamer and M. Silberstein (eds.), *The Blackwell Guide to the Philosophy of Science*, Oxford, Blackwell.

Moore, D. S. 2001, *The Dependent Gene*, New York, W. H. Freeman.

Moss, L. 2003, *What Genes Can't Do*, Cambridge, Mass., MIT Press.

Munsinger, H. 1977, "The Identical-Twin Transfusion Syndrome: A Source of Error in Estimating IQ Resemblance and Heritability," *Annals of Human Genetics* 40: 307–321.

Neale, M. C. and Cardon, L. R. 1992, *Methodology for Genetic Studies of Twins and Families*, Dordrecht, Reidel.

Neisser, U. 1997, "Never a Dull Moment," *American Psychologist* 52: 79–81.

Neisser, U., Boodoo, G., Bouchard, T. J., Jr., et al. 1996, "Intelligence: Knowns and Unknowns," *American Psychologist* 51: 77–101.

Nelkin, D. and Andrews, L. 1996, "The Bell Curve: A Statement," *Science* 271: 13–14.

Nelkin, D. M. and Lindee, S. 1995, *The DNA Mystique: The Gene as a Cultural Icon*, New York, Freeman.

Nichols, R. C. 1987, "Reply to Flynn," in S. Modgil and C. Modgil (eds.), *Arthur Jensen: Consensus and Controversy*, London, Falmer Press.

Nietzsche, F. 1973, *Beyond Good and Evil: Prelude to a Philosophy of the Future*, Harmondsworth, Penguin.

Nisbett, R. E. 1998, "Race, Genetics, and IQ," in C. Jencks and M. Phillips (eds.), *The Black–White Test Score Gap*, Washington, D.C., Brookings Institution Press.

Nuffield 2002, *Genetics and Human Behavior: The Ethical Context*, London, Nuffield Council on Bioethics.

O'Flaherty, B. and Shapiro, J. S. 2004, "Apes, Essences, and Races," in D. Colander, R. E. Prasch, and F. A. Sheth (eds.), *Race, Liberalism, and Economics*, Ann Arbor, University of Michigan Press.

Okasha, S. 2000, Review of S. Sarkar, *Genetics and Reductionism*, *British Journal for the Philosophy of Science* 51: 181–185.

2003, Review of Kitcher's *In Mendel's Mirror, Notre Dame Philosophical Reviews*, September 1, 2003, http://ndpr.nd.edu/review.cfm?id = 1326, accessed August 15, 2004.

Oyama, S. 1988a, "Reply to Robert Plomin's Review of *The Ontogeny of Information*," *Developmental Psychobiology* 21: 97–100.

1988b, "Populations and Phenotypes: A Review of *Development, Genetics and Psychology*," *Developmental Psychobiology* 21: 101–105.

1992, "Ontogeny and Phylogeny: A Case for Metarecapitulation," in P. E. Griffiths (ed.), *Trees of Life*, Dordrecht, Reidel.

2000a, *Evolution's Eye: A Systems View of the Biology–Culture Divide*, Durham, N.C., Duke University Press.

2000b, "Causal Democracy and Causal Contributions in Developmental Systems Theory," *Philosophy of Science* 67: S332–S347.

2000c, *The Ontogeny of Information: Developmental Systems and Evolution*, 2nd edn., Durham, N.C., Duke University Press.

2002, "The Nurturing of Natures," in A. Grunwald, M. Gutmann, and E. M. Neumann-Held (eds.), *On Human Nature: Anthropological, Biological and Philosophical Foundations*, Berlin, Springer.

Oyama, S., Griffiths, P. E., and Gray, R. D. 2001, "Introduction," in S. Oyama, P. E. Griffiths, and R. D. Gray (eds.), *Cycles of Contingency: Developmental Systems and Evolution*, Cambridge, Mass., MIT Press.

Page, E. B. 1972, "Behavior and Heredity," *American Psychologist* 27: 660–661.

Papineau, D. 1982, "Nature and Nurture," *Journal of Medical Ethics* 8: 96–99.

Parens, E. 2004, "Genetic Differences and Human Identities: On Why Talking about Behavior Genetics is Important and Difficult," *Hastings Center Report Special Supplement* 34: S1–S36.

Park, M. A. 2002, *Biological Anthropology*, 3rd edn., New York, McGraw-Hill.

Paul, D. B. 1998, *The Politics of Heredity*, Albany, N.Y., SUNY Press.

Pigliucci, M. 2001, *Phenotypic Plasticity: Beyond Nature and Nurture*, Baltimore, Md., Johns Hopkins University Press.

Pinker, S. 2002, *The Blank Slate: The Modern Denial of Human Nature*, New York, Viking.

Platt, S. A. and Bach, M. 1997, "Use and Misinterpretations of Genetics in Psychology," *Genetica* 99: 135–143.

Platt, S. A. and Stanislow, C. A. 1988, "Norm of Reaction: Definition and Misinterpretation of Animal Research," *Journal of Comparative Psychology* 102: 254–261.

Plomin, R. 1986, *Development, Genetics, and Psychology*, Hillsdale, N.J., Lawrence Erlbaum.

1987, "Genetics of Intelligence," in S. Modgil and C. Modgil (eds.), *Arthur Jensen: Consensus and Controversy*, New York, Falmer Press.

1988, "Reply to Susan Oyama's Review of *Development, Genetics and Psychology*," *Developmental Psychobiology* 21: 107–112.

1990, "Trying to Shoot the Messenger for his Message," *Behavioral and Brain Sciences* 13: 144.

2002, "Quantitative Trait Loci and General Cognitive Ability," in J. Benjamin, R. P. Ebstein, and R. H. Belmaker (eds.), *Molecular Genetics and*

Human Personality, Washington, D.C., American Psychiatric Publishing, Inc.

2003, "Molecular Genetics and *g*," in H. Nyborg (ed.), *The Scientific Study of General Intelligence: Tribute to Arthur R. Jensen*, Amsterdam, Pergamon.

2004, "Intelligence: Genetics, Genes, and Genomics," *Journal of Personality and Social Psychology* 86: 112–129.

Plomin, R. and DeFries, J. C. 1976, Letter, *Science* 194: 9–12.

Plomin, R. and Hershberger, S. 1991, "Genotype–Environment Interaction," in T. D. Wachs and R. Plomin (eds.), *Conceptualization and Measurement of Organism–Environment Interaction*, Washington, D.C., American Psychological Association.

Plomin, R., Willerman, L., and Loehlin, J. C. 1976, "Resemblance in Appearance and the Equal Environments Assumption in Twin Studies of Personality Traits," *Behavior Genetics* 6: 43–52.

Plomin, R., DeFries, J. C., and Loehlin, J. C. 1977, "Genotype–Environment Interaction and Correlation in the Analysis of Human Behavior," *Psychological Bulletin* 84: 309–322.

Plomin, R., DeFries, J. C., and Fulker, D. W. 1988, *Nature and Nurture During Infancy and Early Childhood*, Cambridge, Cambridge University Press.

Plomin, R., DeFries, J. C., McClearn, G. E., and McGuffin, P. 2001, *Behavioral Genetics*, San Francisco, Calif., W. H. Freeman.

Pollak, E., Kempthorne, O., and Bailey, T. B., Jr. (eds.) 1977, *Proceedings of the International Conference on Quantitative Genetics*, Ames, Iowa State University Press.

Price, B. 1950, "Primary Biases in Twin Studies: A Review of Prenatal and Natal Difference-Producing Factors in Monozygotic Pairs," *American Journal of Human Genetics* 2: 293–352.

Provine, W. B. 1986, "Geneticist and Race," *American Zoologist* 26: 857–887.

Putnam, H. 1973, "Reductionism and the Nature of Psychology," *Cognition* 2: 131–146.

Rao, D. C., Morton, N. E., and Yee, S. 1974, "Analysis of Family Resemblance: Linear Model for Familial Correlation," *American Journal of Human Genetics* 26: 331–359.

Redding, R. E. 2001, "Sociopolitical Diversity in Psychology: The Case for Pluralism," *American Psychologist* 56: 205–215.

Relethford, J. H. 2002, "Apportionment of Global Human Genetic Diversity Based on Craniometrics and Skin Color," *American Journal of Physical Anthropology* 118: 393–398.

Richardson, R. C. 1980, Review of M. Sahlins, *Uses and Abuses of Biology*, *Teaching Philosophy* 3: 479–481.

1984, "Biology and Ideology: The Interpenetration of Science and Values," *Philosophy of Science* 51: 396–420.

Ridley, M. 2000, *Genome: The Autobiography of a Species in 23 Chapters*, New York, Harper/Collins.

2003a, *Nature Via Nurture*, New York, Harper/Collins.

2003b, "Listen to the Genome: It's not Nature versus Nurture Anymore," *American Spectator* 36: 59–65.

References

Rijsdijk, F. V., Sham, P. C., Sterne, A., Purcell, S., McGuffin, P., and Plomin, R. 2001, "Life Events and Depression in a Community Sample of Siblings," *Psychological Medicine* 31: 401–410.

Robert, J. S. 2003, "Constant Factors and Hedgeless Hedges: On Heuristics and Biases in Biological Research," *Philosophy of Science* 70: 975–988.

Roberts, R. C. 1967, "Some Concepts and Methods in Quantitative Genetics," in J. Hirsch (ed.), *Behavior-Genetic Analysis*, New York, McGraw-Hill.

Rodgers, J. L. 2000, "A Critique of the Flynn Effect: Massive IQ Gains, Methodological Artifacts, or Both?" *Intelligence* 26: 337–356.

Rose, S., Lewontin, R. C., and Kamin, L. 1984, *Not in Our Genes*, Harmondsworth, Penguin.

Rosenberg, A. 1987, "Is there Really 'Juggling,' 'Artifice,' and 'Trickery' in *Genes, Mind and Culture*," *Behavioral and Brain Sciences* 10: 80–82.

Roughgarden, J. 1979, *Theory of Population Genetics and Evolutionary Ecology: An Introduction*, New York, Macmillan.

Rowe, D. C. 1983, "Biometrical Genetic Models of Self-Reported Delinquent Behavior: A Twin Study," *Behavior Genetics* 13: 473–489.

1993, "Genetic Perspectives on Personality," in R. Plomin and G. E. McClearn (eds.), *Nature, Nurture and Psychology*, Washington, D.C., American Psychological Association.

1994, *The Limits of Family Influence: Genes, Experience, and Behavior*, New York, Guilford Press.

1997, "A Place at the Policy Table? Behavior Genetics and Estimates of Family Environmental Effects on IQ," *Intelligence* 24: 133–158.

2002, *Biology and Crime*, Los Angeles, Calif., Roxbury.

Rowe, D. C. and Rodgers, J. L. 2002, "Expanding Variance and the Case of Historical Changes in IQ Means: A Critique of Dickens and Flynn," *Psychological Review* 109: 759–763.

Rowe, D. C., Vaszony, A. T., and Flannery, D. J. 1994, "No More than Skin Deep: Ethnic and Racial Similarity in Developmental Process," *Psychological Review* 101: 396–413.

1995, "Ethnic and Racial Similarity in Developmental Process: A Study of Academic Achievement," *Psychological Science* 6: 33–38.

Rushton, J. P. 1989, "The Generalizability of Genetic Estimates," *Personality and Individual Differences* 10: 985–989.

1995, *Race, Evolution, and Behavior: A Life History Perspective*, New Brunswick, N.J., Transaction.

Rutter, M. 1997, "Nature–Nurture Integration: The Example of Antisocial Behavior," *American Psychologist* 52: 390–398.

2000, "Closing Remarks," in G. Bock, J. Goode, K. Webb (eds.), *The Nature of Intelligence*, New York, Wiley.

2002, "Nature, Nurture, and Development: From Evangelism through Science toward Policy and Practice," *Child Development* 73: 1–21.

Rutter, M. and Plomin, R. 1997, "Opportunities for Psychiatry from Genetic Findings," *British Journal of Psychiatry* 171: 209–219.

Rutter, M. and Silberg, J. 2002, "Gene–Environment Interplay in Relation to Emotional and Behavioral Disturbance," *Annual Review of Psychology* 53: 463–490.

Ryan, A. 1995, "Apocalypse Now?" in R. Jacoby and N. Glauberman (eds.), *The Bell Curve Debate*, New York, Random House.

Ryle, G. 1974, "Intelligence and the Logic of the Nature–Nurture Issue," *Journal of Philosophy of Education* 8: 52–60.

Salmon, M. H. 1995, *Introduction to Logic and Critical Thinking*, 3rd edn., Fort Worth, Tex., Harcourt, Brace & Co.

Samuelson, F. 1982, "Intelligence and Some of its Testers," *Science* 215: 656–657.

Sarkar, S. 1998, *Genetics and Reductionism*, Cambridge, Cambridge University Press.

1999, "Delusions about IQ," *Cahiers de Psychologie Cognitive* 18: 224–231.

Scarr, S. 1968, "Environmental Bias in Twin Studies," *Eugenics Quarterly* 15: 34–40.

1987, "Three Cheers for Behavior Genetics: Winning the War and Losing our Identity," *Behavior Genetics* 17: 219–228.

1988, "Race and Gender as Psychological Variables," *American Psychologist* 43: 56–59.

Scarr, S. and Carter-Saltzman, L. 1982, "Genetics and Intelligence," in R. J. Sternberg (ed.), *Handbook of Human Intelligence*, Cambridge, Cambridge University Press.

Schaffner, K. F. 1999, "Complexity and Research Strategies in Behavioral Genetics," in R. A. Carson and M. A. Rothstein (eds.), *Behavioral Genetics: The Clash of Culture and Biology*, Baltimore, Md., Johns Hopkins University Press.

2001, "Nature and Nurture," *Current Opinion in Psychiatry* 14: 485–490.

Schiff, M. and Lewontin, R. 1986, *Education and Class: The Irrelevance of IQ Genetic Studies*, Oxford, Clarendon Press.

Schneider, S. M. 2003, "Evolution, Behavior Principles, and Developmental Systems," *Journal of the Experimental Analysis of Behavior* 79: 137–152.

Scriven, M. 1970, "The Values of the Academy: Moral Issues for American Education and Educational Research Arising from the Jensen Case," *Review of Educational Research* 40: 541–549.

Searle, J. 1976, "Rules of the Language Game," *Times Literary Supplement*, September 10.

Segerstrale, U. 2000, *Defenders of the Truth: The Battle for Science in the Sociobiology Debate and Beyond*, Oxford, Oxford University Press.

Sesardic, N. 1993, "Heritability and Causality," *Philosophy of Science* 60: 396–418.

2000, "Philosophy of Science that Ignores Science: Race, IQ and Heritability," *Philosophy of Science* 67: 580–602.

2003, "Heritability and Indirect Causation," *Philosophy of Science* 70: 1002–1014.

Shostak, S. 2003, "Locating Gene–Environment Interaction: At the Intersections of Genetics and Public Health," *Social Science & Medicine* 56: 2327–2342.

Singer, P. 1996, "Ethics and the Limits of Scientific Freedom," *Monist* 79: 218–229.

References

Snyderman, M. and Rothman, S. 1988, *The IQ Controversy: The Media and Public Policy*, New Brunswick, N.J., Transaction Books.

Sober, E. 1984, *The Nature of Selection*, Cambridge, Mass., MIT Press.

1993, *Philosophy of Biology*, Boulder, Colo., Westview Press.

1994, *From a Biological Point of View*, Cambridge, Cambridge University Press.

2000, "The Meaning of Genetic Causation," in A. Buchanan, N. Daniels, D. Wikler, D. W. Brock, and D. I. Wilker (eds.), *From Chance to Choice: Genetics and Justice*, Cambridge, Cambridge University Press.

2001, "Separating Nature and Nurture," in D. Wasserman and R. Wachbroit (eds.), *Genetics and Criminal Behavior*, Cambridge, Cambridge University Press.

Solzhenitsyn, A. 1976, *Warning to the West*, New York, Farrar, Straus and Giroux.

Sparks, C. S. and Jantz, R. L. 2002, "A Reassessment of Human Cranial Plasticity: Boas Revisited," *Proceedings of the National Academy of Sciences* 99: 14636–14639.

Spencer, M. B. and Harpalani, V. 2004, "Nature, Nurture and the Question of 'How,'" in C. G. Coll, E. L. Bearer, and R. M. Lerner (eds.), *Nature and Nurture*, Mahwah, N.J., Lawrence Erlbaum.

Spitz, H. H. (ed.) 1986, *The Raising of Intelligence*, Hillsdale, N.J., Erlbaum.

Sterelny, K. and Griffiths, P. E. 1999, *Sex and Death: An Introduction to Philosophy of Biology*, Chicago, University of Chicago Press.

Stern, B. J. 1950a, "Does Genetic Endowment Vary by Socioeconomic Group?" *Science* 111: 697–698.

Stern, C. 1950b, "Does Genetic Endowment Vary by Socioeconomic Group?" *Science* 111: 698–699.

1973, *Principles of Human Genetics*, 3rd edn., San Francisco, Calif., W. H. Freeman.

Sternberg, R. J. (ed.) 1982, *Handbook of Human Intelligence*, Cambridge, Cambridge University Press.

2003, "'My House Is a Very Very Very Fine House' – But It Is Not the Only House," in H. Nyborg (ed.), *The Scientific Study of General Intelligence: Tribute to Arthur R. Jensen*, Amsterdam, Pergamon.

Sternberg, R. J. and Grigorenko, E. L. 1999, "Myths in Psychology and Education Regarding the Gene–Environment Debate," *Teachers College Record* 100: 536–553.

Stigler, S. M. 1999, *Statistics on the Table: The History of Statistical Concepts and Methods*, Cambridge, Mass., Harvard University Press.

Stoltenberg, S. F. 1997, "Coming to Terms with Heritability," *Genetica* 99: 89–96.

Taylor, K. A. 2001, "On the Explanatory Limits of Behavioral Genetics," in D. Wasserman and R. Wachbroit (eds.), *Genetics and Criminal Behavior*, Cambridge, Cambridge University Press.

Terman, L. M. 1917, "The Intelligence Quotient of Francis Galton in Childhood," *American Journal of Psychology* 28: 209–215.

Thoday, J. M. 1976, "Educability and Group Differences," in N. J. Block and G. Dworkin (eds.), *The IQ Controversy: Critical Readings*, New York, Pantheon.

Thompson, L. A. 1996, "Where Are the Environmental Influences on IQ?" in D. K. Detterman (ed.), *Current Topics in Human Intelligence*, vol. 5: *The Environment*, Norwood, N.J., Ablex.

Thompson, W. R. 1975, "Discussion," *Science* 188: 1125–1126.

Turkheimer, E. 1998, "Heritability and Biological Explanation," *Psychological Review* 105: 782–791.

    2000, "Three Laws of Behavior Genetics and What They Mean," *Current Directions in Psychological Science* 9: 160–164.

Turkheimer, E., Goldsmith, H. H., and Gottesman, I. I. 1995, "Commentary," *Human Development* 38: 142–153.

Tversky, A. and Kahneman, D. 1980, "Causal Schemas in Judgments under Uncertainty," in M. Fishbein (ed.), *Progress in Social Psychology*, Hillsdale, N.J., Erlbaum.

Urbach, P. 1974, "Progress and Degeneration in the 'IQ Debate,'" *British Journal for the Philosophy of Science* 25: 99–135; 235–259.

Venter, C. 2000, "Celera Genomics Remarks at the Human Genome Announcement," http://www.celera.com/celera/pr_1056647999, accessed June 9, 2004.

Vitzthum, V. J. 2003, "A Number No Greater than the Sum of Its Parts: The Use and Abuse of Heritability," *Human Biology* 75: 539–558.

Vonnegut, K. 1970, *Welcome to the Monkey House*, New York, Dell.

Vreeke, G. J. 2000a, "Reply to Turkheimer and Waldron and Wahlsten," *Human Development* 43: 53–56.

    2000b, "Nature, Nurture and the Future of the Analysis of Variance," *Human Development* 43: 32–45.

Wachbroit, D. 2001, "Understanding the Genetics-of-Violence Controversy," in D. Wasserman and R. Wachbroit (eds.), *Genetics and Criminal Behavior*, Cambridge, Cambridge University Press.

Waddington, C. H. 1957, *The Strategy of the Genes*, London, George Allen & Unwin.

Wade, N. 2002, "A New Look at Old Data May Discredit a Theory on Race," *New York Times*, October 8.

Wahlsten, D. 1990a, "Insensitivity of the Analysis of Variance to Heredity–Environment Interaction," *Behavioral and Brain Sciences* 13: 109–120.

    1990b, "Goals and Methods: The Study of Development versus Partitioning of Variance," *Behavioral and Brain Sciences* 13: 146–155.

    1994, "Nascent Doubts May Presage Conceptual Clarity: Reply to Surbey," *Canadian Journal of Psychology* 35: 265–267.

    1995, "The Intelligence of Heritability," *Canadian Psychology* 35: 244–258.

    1997, "The Malleability of Intelligence is not Constrained by Heritability," in B. Devlin, S. Fienberg, D. P. Resnick, and K. Roeder (eds.), *Intelligence, Genes, and Success: Scientists Respond to* The Bell Curve, New York, Springer.

    2000a, "Analysis of Variance in the Service of Interactionism," *Human Development* 43: 46–50.

    2000b, "Behavioral Genetics," in A. E. Kazdin (ed.) *Encyclopedia of Psychology*, Oxford, Oxford University Press.

2003, "Genetics and the Development of Brain and Behavior," in J. Valsiner and K. J. Connoly (eds.), *Handbook of Developmental Psychology*, London, Sage.

Wahlsten, D. and Gottlieb, G. 1997, "The Invalid Separation of Effects of Nature and Nurture: Lessons from Animal Experimentation," in R. J. Sternberg and E. Grigorenko (eds.), *Intelligence, Heredity and Environment*, New York, Cambridge University Press.

Waldman, I. 1997, "Unresolved Questions and Future Directions in Behavior-Genetic Studies of Intelligence," in R. J. Sternberg and E. Grigorenko, (eds.), *Intelligence, Heredity, and Environment*, New York, Cambridge University Press.

Ward, J. 1998, "Sir Cyril Burt: The Continuing Saga," *Educational Psychology* 18: 235–242.

Wasserman, D. 2004, "Is There Value in Identifying Individual Genetic Predispositions to Violence?" *Journal of Law, Medicine and Ethics* 32: 24–33.

Watson, J. D. 1986, "Biology: A Necessarily Limitless Vista," in S. Rose and L. Appignanesi (eds.), *Science and Beyond*, Oxford, Blackwell.

2003, "A Molecular Genetic Perspective," in R. Plomin, J. C. DeFries, I. W. Craig, and P. McGuffin (eds.), *Behavioral Genetics in the Postgenomic Era*, Washington, D.C., American Psychological Association.

Weinberg, S. 1993, *Dreams of a Final Theory*, London, Hutchinson.

West-Eberhard, M. J. 2003, *Developmental Plasticity and Evolution*, Oxford, Oxford University Press.

Wilson, E. O. 1977, "In Response to 'Unnatural Science,'" *New York Review of Books*, March 31.

Worrall, J. 2000, "Pragmatic Factors in Theory Acceptance," in W. H. Newton-Smith (ed.) *A Companion to the Philosophy of Science*, Oxford, Blackwell.

Index

Milton Keynes UK
Ingram Content Group UK Ltd.
UKHW041523181024
449640UK00009B/163

9 780521 173339